Master Math: Calculus

By
Debra Anne Ross

Course Technology PTR

A part of Cengage Learning

COURSE TECHNOLOGY
CENGAGE Learning™

Australia • Brazil • Japan • Korea • Mexico • Singapore • Spain • United Kingdom • United States

COURSE TECHNOLOGY
CENGAGE Learning™

Publisher and General Manager, Course Technology PTR: Stacy L. Hiquet

Associate Director of Marketing: Sarah Panella

Manager of Editorial Services: Heather Talbot

Marketing Manager: Jordan Casey

Senior Acquisitions Editor: Emi Smith

Interior Layout Tech: Judith Littlefield

Illustrations and Equations: Judith Littlefield

Cover Designer: Jeff Cooper

Indexer: Larry Sweazy

Proofreader: Jenny Davidson

For product information and technology assistance, contact us at **Cengage Learning Customer & Sales Support, 1-800-354-9706**
For permission to use material from this text or product, submit all requests online at **cengage.com/permissions**
Further permissions questions can be emailed to **permissionrequest@cengage.com**

All trademarks are the property of their respective owners.

Library of Congress Control Number: 2009924627

ISBN-10: 1-59863-986-2

ISBN-13: 978-1-59863-986-5

Course Technology, a part of Cengage Learning
20 Channel Center Street
Boston, MA 02210
USA

Cengage Learning is a leading provider of customized learning solutions with office locations around the globe, including Singapore, the United Kingdom, Australia, Mexico, Brazil, and Japan. Locate your local office at: **international.cengage.com/region**

Cengage Learning products are represented in Canada by Nelson Education, Ltd. For your lifelong learning solutions, visit **courseptr.com**
Visit our corporate website at **cengage.com**

Printed in Canada
1 2 3 4 5 6 7 12 11 10 09

Table of Contents

Introduction   **1**

Chapter 1: Functions   **3**

1.1   Functions: Types, Properties, and Definitions   3

1.2   Exponents and Logarithms   8

1.3   Trigonometric Functions   10

1.4   Circular Motion   20

1.5   Relationship Between Trigonometric and
       Exponential Functions   25

1.6   Hyperbolic Functions   26

1.7   Polynomial Functions   28

1.8   Functions of More Than One Variable and Contour Diagrams   29

1.9   Coordinate Systems   33

1.10   Complex Numbers   36

1.11   Parabolas, Circles, Ellipses, and Hyperbolas   38

Chapter 2: The Derivative   **47**

2.1   The Limit   47

2.2   Continuity   50

2.3   Differentiability   52

2.4   The Definition of the Derivative and Rate of Change   53

2.5 Δ (delta) Notation and the Definition of the Derivative   57

2.6   Slope of a Tangent Line and the Definition of the Derivative   58

2.7   Velocity, Distance, Slope, Area, and the Definition
       of the Derivative   60

2.8   Evaluating Derivatives of Constants and Linear Functions   63

2.9   Evaluating Derivatives Using the Derivative Formula   64

2.10   The Derivatives of a Variable, a Constant with a Variable,
       a Constant with a Function, and a Variable Raised to a Power   66

2.11   Examples of Differentiating Using the Derivative Formula   68

2.12   Derivatives of Powers of Functions   69

2.13   Derivatives of a^x, e^x, and ln x   71

2.14   Applications of Exponential Equations   77

2.15   Differentiating Sums, Differences, and Polynomials   80

2.16   Taking Second Derivatives   81

2.17   Derivatives of Products: The Product Rule   82

2.18   Derivatives of Quotients: The Quotient Rule   85

2.19   The Chain Rule for Differentiating Complicated Functions   86

2.20   Rate Problem Examples   90

2.21   Differentiating Trigonometric Functions   91

2.22   Inverse Functions and Inverse Trigonometric Functions
       and Their Derivatives   95

2.23   Differentiating Hyperbolic Functions   99

2.24   Differentiating Multivariable Functions   101

2.25   Differentiation of Implicit Vs. Explicit Functions   101

2.26   Selected Rules of Differentiation   102

2.27   Minimum, Maximum, and the First and Second Derivatives   103

2.28   Notes on Local Linearity, Approximating Slope of Curve,
       and Numerical Methods   109

Chapter 3: The Integral   **113**

3.1   Introduction   113

3.2   Sums and Sigma Notation   114

3.3   The Antiderivative or Indefinite Integral and the
       Integral Formula   117

3.4   The Definite Integral and the Fundamental Theorem
       of Calculus   120

3.5   Improper Integrals   122

3.6   The Integral and the Area Under a Curve   124

3.7   Estimating Integrals Using Sums and the Associated Error   128

3.8   The Integral and the Average Value   131

3.9   Area Below the X-axis, Even and Odd Functions, and
       The Integrals   131

3.10   Integrating a Function and a Constant, the Sum of Functions,
       a Polynomial, and Properties of Integrals   134

3.11   Multiple Integrals   136

3.12   Examples of Common Integrals   138

3.13   Integrals Describing Length   139

3.14   Integrals Describing Area   140

3.15   Integrals Describing Volume   145

3.16   Changing Coordinates and Variables   152

3.17   Applications of the Integral   157

3.18   Evaluating Integrals Using Integration by Parts   162

3.19   Evaluating Integrals Using Substitution   164

3.20   Evaluating Integrals Using Partial Fractions   172

3.21   Evaluating Integrals Using Tables   177

Chapter 4: Series and Approximation   **179**

4.1   Sequences, Progressions, and Series   179

4.2   Infinite Series and Tests for Convergence   183

4.3   Expanding Functions Into Series, the Power Series,
       Taylor Series, Maclaurin Series, and the Binomial Expansion   188

Chapter 5: Vectors, Matrices, Curves, Surfaces, and Motion   **195**

5.1   Introduction to Vectors   195

5.2   Introduction to Matrices   202

5.3   Multiplication of Vectors and Matrices   205

5.4   Dot or Scalar Products   208

5.5   Vector or Cross Product   211

5.6   Summary of Determinants   215

5.7   Matrices and Linear Algebra   217

5.8   The Position Vector Parametric Equations, Curves,
       and Surfaces   224

5.9   Motion, Velocity, and Acceleration   230

Chapter 6: Partial Derivatives   **243**

6.1   Partial Derivatives: Representation and Evaluation   243

6.2   The Chain Rule   246

6.3   Representation on a Graph   247

6.4   Local Linearity, Linear Approximations, Quadratic
      Approximations, and Differentials   250

6.5   Directional Derivative and Gradient   255

6.6   Minima, Maxima, and Optimization   259

Chapter 7: Vector Calculus   **267**

7.1   Summary of Scalars, Vectors, the Directional Derivative,
      and the Gradient   267

7.2   Vector Fields and Field Lines   271

7.3   Line Integrals and Conservative Vector Fields   276

7.4   Green's Theorem: Tangent and Normal (Flux) Forms   282

7.5   Surface Integrals and Flux   287

7.6   Divergence   295

7.7   Curl   300

7.8   Stokes' Theorem   304

Chapter 8: Introduction to Differential Equations   **307**

8.1   First-Order Differential Equations   308

8.2   Second-Order Linear Differential Equations   312

8.3   Higher-Order Linear Differential Equations   315

8.4   Series Solutions to Differential Equations   317

8.5   Systems of Differential Equations   319

8.6   Laplace Transform Method   321

8.7   Numerical Methods for Solving Differential Equations   322

8.8   Partial Differential Equations   324

Acknowledgments

I am indebted to Dr. Cyndy Lakowske for reading this book for accuracy and for all of her helpful comments. I am also indebted to Dr. Melanie McNeil, Professor of Chemical Engineering at San Jose State University, for reading the *Master Math* books and for all of her helpful comments. I am grateful to Dr. Channing Robertson, Professor of Chemical Engineering at Stanford University, for reviewing this book and, in general, for all his guidance. I especially thank my mother, Maggie Ross, for her editorial help.

Without my wonderful agent, Sidney B. Kramer, and the staff of Mews Books, the *Master Math* series would not have been published. Thank you, Sidney! I am also thankful to Ron Fry and the staff of Career Press for their work in publishing and launching the original *Master Math* books as a successful series.

I am grateful to Emi Smith, Senior Acquisitions Editor, and Course Technology, a part of Cengage Learning, for invigorating the *Master Math* series and improving the presentation. I particularly appreciate Judith Littlefield for her tireless and expert work on the illustrations, equations, and layout. Much thanks to Jenny Davidson for proofreading, Jeff Cooper for cover design, Larry Sweazy for indexing, as well as Stacy L. Hiquet, Sarah Panella, Heather Talbot, and Jordan Casey.

Finally, I deeply appreciate my beautiful and brilliant husband, David A. Lawrence, who worked side-by-side with me as we meticulously edited text and figures.

About the Author

Debra Anne Ross Lawrence is the author of six books of the *Master Math* series: *Basic Math and Pre-Algebra*, *Algebra*, *Pre-Calculus*, *Calculus*, *Trigonometry*, and *Geometry*. She earned a double Bachelor of Arts degree in biology and chemistry with honors from the University of California at Santa Cruz and a Master of Science degree in chemical engineering from Stanford University.

Her research experience encompasses investigating the photosynthetic light reactions using a dye laser, studying the eye lens of diabetic patients, creating a computer simulation program of physiological responses to sensory and chemical disturbances, genetically engineering bacteria cells for over-expression of a protein, and designing and fabricating biological reactors for in-vivo study of microbial metabolism using nuclear magnetic resonance spectroscopy.

Debra was a member of a small team of scientists and engineers who developed and brought to market the first commercial biosensor system. She managed an engineering group responsible for scale-up of combinatorial synthesis for pharmaceutical development. She also managed intellectual property for a scientific research and development company. Debra's work has been published in scientific journals and/or patented.

Debra is also the author of *The 3:00 PM Secret: Live Slim and Strong Live Your Dreams* and *The 3:00 PM Secret 10-Day Dream Diet*. She is the coauthor with her husband, David A. Lawrence, of *Arrows Through Time: A Time Travel Tale of Adventure, Courage, and Faith*. Debra is President of GlacierDog Publishing and Founder of GlacierDog.com. When Debra is not engaged in all-season mountaineering near her Alaska home, she is endeavoring to understand the incomprehensible workings of the universe.

Introduction

Master Math: Calculus is a comprehensive reference book for advanced high school and college students that explains and clarifies the key principles of calculus. The purpose of the book is to provide an easy-to-access reference source for locating specific calculus topics. This book is designed so that a student can quickly look up a topic and, by reading the explanation and the information in its section, find the relevant facts and formulas. This book can also be used to obtain a general knowledge and understanding of calculus and it provides a complete breadth of material so that most topics related to calculus are explained.

Master Math: Calculus reviews functions and explains the principles and operations of the derivative, the integral, series and approximations, vectors, matrices, curves, surfaces, motion, partial derivatives, vector calculus, and introductory differential equations. The chapters in this book are divided into major sections containing independent topics housed within the context of where they fit into the discipline of calculus.

The *Master Math* series includes *Basic Math and Pre-Algebra, Algebra, Pre-Calculus, Calculus, Geometry,* and *Trigonometry.* The *Master Math* series presents the general principles of mathematics from grade school through college including arithmetic, algebra, geometry, trigonometry, pre-calculus, and calculus.

Chapter

1

Functions

1.1 Functions: Types, Properties, and Definitions

- This section includes definitions and explanations of functions, domain set, range set, graphing functions, compound functions, inverse functions, as well as adding, subtracting, multiplying and dividing functions, linear and non-linear functions, and even and odd functions.

- *Functions* are an integral part of calculus. Functions reflect the fact that one or more properties can depend on (are a function of) another property. For example, how fast a trolley cart can carry a rock up a hill is a function of how much the rock weighs, the slope of the hill, and the horsepower of the motor.

- Common functions used in calculus include trigonometric functions, logarithmic functions, and exponential functions. (See *Master Math: Pre-Calculus* Chapters 2, 3, and 4 for additional information about functions.)

- A *function* is a relation, rule, expression, or equation that associates each element of a *domain set* with its corresponding element in the *range set*. For a relation, rule, expression, or equation to be a function, there must be only one element or number in the range set for each element or number in the domain set. The domain set of a function is the set of possible values of the independent variable, and the range set is the corresponding set of values of the dependent variable.

• The domain set is the initial set and the range set is the set that results after a function is applied.

domain set → function f( ) → range set

For example:

   domain set x = {2, 3, 4}
   through function f(x) = x^2, f(2) = 2^2, f(3) = 3^2, f(4) = 4^2
   to range set f(x) = {4, 9, 16}.

• The domain set and range set can be expressed as (x,f(x)) pairs. In the previous example, the function is f(x) = x^2 and the pairs are (2,4), (3,9), and (4,16).

• For each member of the domain set, there must be only one corresponding member in the range set. For example:

   F = (2,4), (3,9), (4,16)   where F is a function.
   M = (2,5), (2,–5), (4,9)   where M is not a function.

M is not a function because the number 2 in the domain set corresponds to more than one number in the range set.

• Functions can be expressed in the form of a graph, a formula, or a table. To *graph functions*, the values in the domain set correspond to the X-axis and the related values in the range set correspond to the Y-axis. For example:

   domain set x = –2, –1, 0, 2
   through function f(x) = x + 1
   to range set f(x) = –1, 0, 1, 3
   resulting in pairs (x,y) = (–2,–1), (–1,0), (0,1), (2,3).

When graphed these resulting pairs are depicted as:

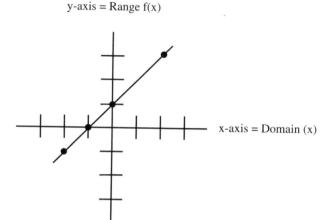

y-axis = Range f(x)

x-axis = Domain (x)

- Graphs of functions only have one value of y for each x value:

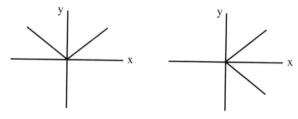

Graph is a function   Graph is not a function

If a vertical line can be drawn that passes through the function more than one time, there is more than one y value for a given x value and the graph is not a function. This is called the *vertical line test*.

- The following are examples of (a.) *addition*, (b.) *subtraction*, (c.) *multiplication*, and (d.) *division* of *functions*. In these examples the functions f(x) and g(x) are given by $f(x) = 2x$ and $g(x) = x^2$:

(a.) $f(x) + g(x) = (f + g)(x) = 2x + x^2$
(b.) $f(x) - g(x) = (f - g)(x) = 2x - x^2$
(c.) $f(x) \times g(x) = (f \times g)(x) = 2x \times x^2 = 2x^3$
(d.) $f(x) \div g(x) = (f \div g)(x) = 2x \div x^2 = (2x)/x^2 = 2/x$

• *Composite or compound functions* are functions that are combined, and the operations specified by the functions are combined. Compound functions are written f(g(x)) or g(f(x)) where there is a function of a function. (See Section 2.19 for differentiating compound functions.)

For example, if f(x) = x + 1 and g(x) = 2x – 2, then the compound functions for f(g(x)) and g(f(x)) are:

f(g(x)) = f(2x – 2) = (2x – 2) + 1 = 2x – 1 and

g(f(x)) = g(x + 1) = 2(x + 1) – 2 = 2x + 2 – 2 = 2x

• *Inverse functions* are functions that result in the same value of x after the operations of the two functions are performed. In inverse functions, the operations of each function are the reverse of the other function. Notation for inverse functions is f⁻¹(x). If f(x) = y, then f⁻¹(y) = x. If f is the inverse of g, then g is the inverse of f. A function has an inverse if its graph intersects any horizontal line no more than once. (Please see the beginning of Section 2.22 for a more complete explanation of inverse functions.)

• Functions can be *linear* or *non-linear*. Remember, *linear equations* are equations in which the variables do not have any exponents other than 1. These equations, if plotted, will produce a straight line. A general form of a linear equation is Ax + By = C, where A, B, and C are constants, and x and y are variables. Another general form of a linear equation is y = mx + b, where m is the slope of the line and b is where the line intercepts the Y-axis on a coordinate system. The equation for the slope of a line passing through point (x_1, y_1) can be written $y - y_1 = m(x - x_1)$.

A *linear function* can have the form y = f(x) = b + mx, where m is the slope of the line and represents the rate of change of y with respect to x, and b is the vertical intercept where the line intercepts the Y-axis on a coordinate system that is the value of y when x equals zero.

The *slope of a linear function* can be calculated using the following equation and the values of the function at two points on the graph of the function at $(x_1, f(x_1))$ and $(x_2, f(x_2))$:

$$f(x_2) - f(x_1) = m(x_2 - x_1) \qquad \text{where m is the slope.}$$

This equation can be equivalently written:

$$m = \frac{f(x_2) - f(x_1)}{x_2 - x_1}$$

The quantity $(f(x_2) - f(x_1))/(x_2 - x_1)$ is the quotient of the two differences and is referred to as a *difference quotient*.

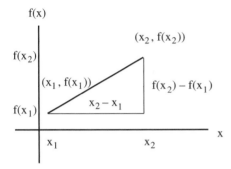

• *Non-linear functions* have variables with exponents greater than 1. Remember that *non-linear equations* are equations containing variables that have exponents greater than 1. Graphs on *non-linear functions* form curved lines and surfaces.

• In general, a *function is increasing* when $y = f(x)$ increases as x increases, and a *function is decreasing* when $y = f(x)$ decreases as x increases.

• A function can be an *even function* or an *odd function*. By determining whether a function is even or odd, it is sometimes possible to simplify an integral of the function to a more manageable form and solve it using symmetry. A function is even if $f(x) = f(-x)$ for all x, and a function is odd if $f(x) = -f(-x)$ for all x.

Examples of *even functions* include:

$$f(x) = c, \, f(x) = x^2, \, f(x) = x^4, \, f(x) = x^{2n}$$
$$f((-x)^2) = (-x)(-x) = x^2$$

cosine is an even function, $\cos(-x) = \cos x$

Examples of odd functions include:

$$f(x) = x, f(x) = x^3, f(x) = x^5, f(x) = x^{2n+1}$$
$$f((-x)^3) = (-x)(-x)(-x) = (-x)^3$$

sine is an odd function, $\sin(-x) = -\sin x$

• By observing the graph of a function, it is clear whether the function is even or odd. If the area between the curve and the X-axis on the section of the function to the left of the Y-axis is equivalent to the area between the curve and the X-axis on the section to the right of the Y-axis, the function is *even*. Therefore, in an even function, the area for negative values along the X-axis is equal to the area for positive values along X-axis.

Alternatively, if a function is *odd*, the area between the curve and the X-axis on the section of the function to the left of the Y-axis is equivalent but opposite to the area between the curve and the X-axis on the section to the right of the Y-axis. Therefore, in an odd function, the area for negative values along the X-axis is equal but opposite to the area for positive values along the X-axis, and the two areas subtract and cancel each other out (which results in the integral being equivalent to zero).

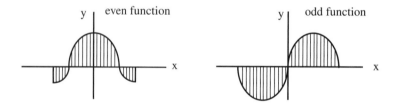

1.2. Exponents and Logarithms

• This section includes exponential functions, logarithms, natural logarithms, and base changes.

• *Exponential functions* form curved lines and contain variables in their exponents. Examples of exponential functions include e^x, a^x, and 2^x, where a is a constant. Some properties of e^x or a^x include:

$$e^x e^y = e^{x+y}$$
$$e^x/e^y = e^{x-y}$$
$$(e^x)^y = e^{xy}$$
$$e^0 = 1$$
$$e = 2.71828\ 18284\ 59045\ 23536\ 02874\ 71353$$

• The inverse of e^x is ln x or the *natural logarithm* of x. Some properties of ln x include:

$$\ln(xy) = \ln x + \ln y$$
$$\ln(x/y) = \ln x - \ln y$$
$$\ln x^y = y \ln x$$
$$\ln(e^x) = x$$
$$e^{\ln x} = x$$
$$e^{-\ln x} = e^{\ln(1/x)} = 1/x$$
$$\ln x = \log_e x = (2.3026)\log x$$

• *Logarithms* can have any base. Base 10 logarithms are the most common and are written $\log_{10} x$ or just log x. The inverse of log x is 10^x. Some properties of log x include:

$$10^{\log x} = x$$
$$10^{-\log x} = 1/x$$
$$\log(xy) = \log x + \log y$$
$$\log(x/y) = \log x - \log y$$
$$\log x^y = y \log x$$
$$\log(10^x) = x$$

• It is important to remember that when a number has an exponent, the logarithm is the exponent. For example:

$$\log(10^x) = x$$
$$\ln(e^x) = x$$
$$\log(10^3) = 3$$
$$\log(10^{-2}) = \log(1/10^2) = -2$$
$$\log(b^x) = x \qquad \text{where b represents any base.}$$

• These principles can be used to solve exponents and logarithms. For example, to solve a base 5 logarithm,

$\log_5(x + 1) = 2$ for x, raise both sides by base 5:
$$5^{\log 5(x + 1)} = 5^2$$
$$(x + 1) = 25$$
$$x = 24$$

- To *change from one base to another*, the following rules apply.
For changing from base b to base a:

$$b = a^{(\log_a b)}$$

$$b^x = a^{(\log_a b)x}$$

If $y = b^x$, then $\log_a y = (\log_a b)x$, or

$$\log_a y = (\log_a b)(\log_b y)$$

- The exponential function e^x and the natural logarithm $\ln x$ can be depicted as:

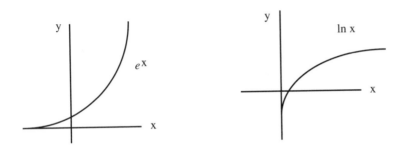

See *Master Math: Basic Math and Pre-Algebra* Chapters 7 and 8 for additional information on exponents and logarithms.

1.3 Trigonometric Functions

- This section includes trigonometric functions, circles, degrees, radians, arc length, Pythagorean formula, distance between two points, addition and subtraction formulas, double angle formulas, graphs of trigonometric functions, and inverse trigonometric functions.

- *Trigonometric functions* can be defined using ratios of sides of a right triangle and, more generally, using the coordinates of points on a circle of radius one. Trigonometric functions are sometimes called *circular functions* because their domains are lengths of arcs on a circle. *Sine, cosine, tangent, cotangent, secant,* and *cosecant* are trigonometric functions.

• The three sides of a *triangle* provide six important trigonometric functions:

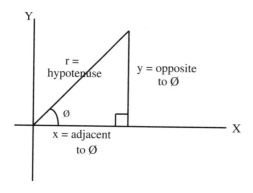

sine $\emptyset$ = sin $\emptyset$ = y/r

cosine $\emptyset$ = cos $\emptyset$ = x/r

tangent $\emptyset$ = tan $\emptyset$ = y/x = sin $\emptyset$ / cos $\emptyset$

cosecant $\emptyset$ = csc $\emptyset$ = r/y

secant $\emptyset$ = sec $\emptyset$ = r/x = 1/ cos $\emptyset$

cotangent $\emptyset$ = cot $\emptyset$ = x/y = 1/ tan $\emptyset$

• Trigonometric functions can be defined using a *circle* having a radius of one, which describes their *periodic* nature. A circle having a radius of one and a point P with coordinates defined by the angle of the arc formed from the X-axis is:

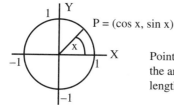

P = (cos x, sin x)

Point P has coordinates (cos x, sin x), and the arc distance of the angle has x units of length and is measured in radians.

The distance around the circle is 2πr and the distance to any point on the circle is defined by the length of the *arc* from the X-axis and the point on the circle that is given by the angle x multiplied by the radius r, or xr. Angles are measured from the positive X-axis in a counterclockwise fashion. Angles measured clockwise are negative.

- In a *circle* having a radius of one, a *radian* is equal to the angle at the center that cuts across an arc of length 1. Remember, a radian is the angle at the center of a circle equal to $57.2957795131°$ which subtends (is opposite to) the *arc* of the circle equal in length to the *radius*. Note the following:

    1 radian $= 360°/2\pi = 180°/\pi = 57.2957795131°$

    2π radians $= 360$ degrees

    $1° = 0.017453292519943$ radian

- An *arc* is a section of a circle defined by two or more points, and can be measured in *degrees* or radians. The following are equivalent to arc length:

    Arc length $=$ (radius)$\times$(central angle measure in radians)
    $= (n°/360°)(\pi d) = r\phi$
    where $n°$ and ϕ represent the *central angle*.

- When angles are measured in *radians*, sin x and cos x have *period* 2π. For each point around a circle the six functions, cos x, sin x, tan x, csc x, sec x, and cot x, can be drawn as six graphs of the corresponding *waveforms*.

- *Pythagorean formula* $x^2 + y^2 = r^2$, can be used to provide many useful formulas. For example, dividing by r^2 gives $(x/r)^2 + (y/r)^2 = 1$, or equivalently $\cos^2\phi + \sin^2\phi = 1$.

Dividing Pythagorean formula by x^2 gives $1 + (y/x)^2 = (r/x)^2$, or equivalently $1 + \tan^2\phi + \sec^2\phi$.

Dividing Pythagorean formula by y^2 gives $(x/y)^2 + 1 = (r/y)^2$, or equivalently $\cot^2\phi + 1 = \csc^2\phi$.

- Pythagorean formula can also be used to calculate the *distance between points*. These points can be defined by X and Y axes of a coordinate system. The distance d between the points is represented using:
$d^2 = (x_2 - x_1)^2 + (y_2 - y_1)^2$ and depicted by:

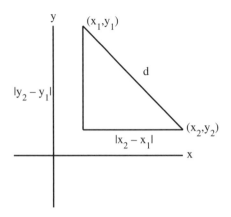

Note that the distance d between two points in three-dimensional space is represented using:

$$d^2 = (x_2 - x_1)^2 + (y_2 - y_1)^2 + (z_2 - z_1)^2$$

• To measure *distance between two points on a circle*, define each point using $P = (\cos \phi, \sin \phi)$ and $P = (\cos \Omega, \sin \Omega)$, where the angle's ϕ and Ω represent the length of the arc they form from the X-axis.

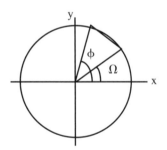

The point on the circle defined by the angle ϕ is at $x = \cos \phi$, $y = \sin \phi$, and the point on the circle defined by Ω is at $x = \cos \Omega$ and $y = \sin \Omega$. The angle between the two points on the circle is $\phi - \Omega$. If (x_1, y_1) and (x_2, y_2) represent the two points, the distance d between the points can be represented by:

$$d^2 = (x_2 - x_1)^2 + (y_2 - y_1)^2$$

This can also be written:

$$d^2 = (\cos \Omega - \cos \phi)^2 + (\sin \Omega - \sin \phi)^2$$
$$= (\cos \Omega - \cos \phi)(\cos \Omega - \cos \phi) + (\sin \Omega - \sin \phi)(\sin \Omega - \sin \phi)$$
$$= \cos^2\Omega - 2 \cos \Omega \cos \phi + \cos^2\phi + \sin^2\Omega - 2 \sin \Omega \sin \phi + \sin^2\phi$$

Because, $\cos^2 + \sin^2 = 1$, d^2 becomes:

$$d^2 = 1 + 1 - 2 \cos \Omega \cos \phi - 2 \sin \Omega \sin \phi$$
$$d^2 = 2 - 2 \cos \Omega \cos \phi - 2 \sin \Omega \sin \phi$$

If the triangle is rotated, the distance between the two points remains the same, but can be represented as:

$$d^2 = (\cos(\Omega - \phi) - 1)^2 + (\sin(\Omega - \phi))^2$$
$$d^2 = 2 - 2 \cos(\Omega - \phi)$$

• Important formulas used in calculus include the addition and subtraction formulas for cosine and sine. The addition and subtraction formulas for cosine can be derived using the fact that the distance between two points on a circle is the same whether a triangle between the two points is rotated or not. Setting the two d^2 equations equal gives:

$$d^2 = 2 - 2 \cos \Omega \cos \phi - 2 \sin \Omega \sin \phi = 2 - 2 \cos(\Omega - \phi)$$
$$= \cos \Omega \cos \phi + \sin \Omega \sin \phi = \cos(\Omega - \phi)$$

Therefore the *subtraction formula for cosine* is:

$$\cos(\Omega - \phi) = \cos \Omega \cos \phi + \sin \Omega \sin \phi$$

Similarly, to obtain the *addition formula for cosine*, $\cos(\Omega + \phi)$, replace ϕ by $(-\phi)$. Therefore:

$$\cos(\Omega + \phi) = \cos \Omega \cos \phi - \sin \Omega \sin \phi$$

These two formulas are known as the addition and subtraction formulas for $\cos(\Omega + \phi)$ and $\cos(\Omega - \phi)$.

• The addition and subtraction formulas for sine can be derived using a right triangle. Right triangles have complimentary angles that can be measured by $90° - \phi$ or $\pi/2 - \phi$.

Therefore:

$$\cos \phi = \sin(\pi/2 - \phi)$$
$$\sin \phi = \cos(\pi/2 - \phi)$$

and,

$$\sin(\Omega + \phi) = \cos(\pi/2 - \Omega - \phi)$$

Using the cosine subtraction formula from above:

$$\cos(\Omega - \phi) = \cos \Omega \cos \phi + \sin \Omega \sin \phi$$

This can be rewritten as:

$$\cos(\pi/2 - \Omega - \phi) = \cos(\pi/2 - \Omega) \cos \phi + \sin(\pi/2 - \Omega) \sin \phi$$

Substituting gives the *addition formula for sine*:

$$\sin(\Omega + \phi) = \sin \Omega \cos \phi + \cos \Omega \sin \phi$$

Similarly, for the subtraction formula for sine, begin with:

$$\cos(\Omega + \phi) = \cos \Omega \cos \phi - \sin \Omega \sin \phi$$
$$\cos(\pi/2 - \Omega + \phi) = \cos(\pi/2 - \Omega) \cos \phi - \sin(\pi/2 - \Omega) \sin \phi$$

Substituting results in the *subtraction formula for sine*:

$$\sin(\Omega - \phi) = \sin \Omega \cos \phi - \cos \Omega \sin \phi$$

• Another important formula is the *double angle formula*, which represents the case when $\Omega = \phi$, so that $\cos(\phi + \phi)$ becomes:

$$\cos(\phi + \phi) = \cos \phi \cos \phi - \sin \phi \sin \phi = \cos 2\phi = \cos^2\phi - \sin^2\phi$$

Substituting $\cos^2\phi + \sin^2\phi = 1$ results in the double angle formula:

$$\cos 2\phi = \cos^2\phi - \sin^2\phi$$
$$\cos 2\phi = (1 - \sin^2 \phi) - \sin^2 \phi = 1 - 2 \sin^2 \phi$$
$$\text{or, } \cos 2\phi = \cos^2 \phi - (1 - \cos^2 \phi) = 2 \cos^2 \phi - 1$$

Similarly, for sine when $\Omega = \phi$, the double angle formula is:

$$\sin(\phi + \phi) = \sin \phi \cos \phi + \cos \phi \sin \phi = \sin 2\phi = 2 \sin \phi \cos \phi$$

- Following are important *trigonometric functions and relations*:

  $\tan x = \sin x / \cos x = 1 / \cot x$

  $\cot x = \cos x / \sin x = 1 / \tan x = \cos x \csc x$

  $\sec x = 1 / \cos x$

  $\csc x = 1 / \sin x$

  $\sin(\pi - x) = \sin x$

  $\cos(\pi - x) = -\cos x$

  $\sin x = \cos(x - \pi/2) = \cos(\pi/2 - x)$

  $\cos x = \sin(x + \pi/2) = \sin(\pi/2 - x)$

  $\sin^2 x + \cos^2 x = 1$

  $1 + \tan^2 x = \sec^2 x$

  $1 + \cot^2 x = \csc^2 x$

  $\sin 2x = 2 \sin x \cos x$

  $\cos 2x = \cos^2 x - \sin^2 x = 2 \cos^2 x - 1 = 1 - 2 \sin^2 x$

  $\sin(x + y) = \sin x \cos y + \cos x \sin y$

  $\sin(x - y) = \sin x \cos y - \cos x \sin y$

  $\cos(x + y) = \cos x \cos y - \sin x \sin y$

  $\cos(x - y) = \cos x \cos y + \sin x \sin y$

  $\sin x \cos y = (1/2)\sin(x - y) + (1/2)\sin(x + y)$

  $\cos x \cos y = (1/2)\cos(x - y) + (1/2)\cos(x + y)$

  $\sin x \sin y = (1/2)\cos(x - y) - (1/2)\cos(x + y)$

  $\tan(x + y) = (\tan x + \tan y) / (1 - \tan x \tan y)$

- *Graphs of trigonometric functions* can be sketched by selecting values for x, calculating the corresponding y values and plotting the curves. Graphs can also be drawn by entering an equation into a graphing calculator or graphing software. If there are coefficients in the equations for $y = \cos x$, $y = \sin x$, etc., the graph of the function will have the same general shape, but it will have a larger or smaller amplitude, or it will be elongated or narrower, or it will be moved to the right or left or up or down.

For example, if there is a coefficient of 2 in front of cosine or sine, the graph will go to +2 and –2 (rather than +1 and –1) on the Y-axis. Similarly, if there is a coefficient of 1/2 in front of cosine or sine, the graph will go to +1/2 and –1/2 (rather than +1 and –1) on the Y-axis.

If, for example, there is a coefficient of 2 in front of x, resulting in $y = \cos 2x$ and $y = \sin 2x$, the graph will complete each cycle along the X-axis twice as fast. Because there is one cycle between 0 and 2π for

y = cos x and y = sin x, there will be two cycles between 0 and 2π for
y = cos 2x and y = sin 2x. Similarly, if there is a coefficient of 1/2 in
front of x, giving y = cos x/2 and y = sin x/2, the graph will complete
each cycle along the X-axis half as fast. Because there is one cycle
between 0 and 2π for y = cos x and y = sin x, there will be one-half
of a cycle between 0 and 2π for y = cos x/2 and y = sin x/2.

Also, if a number is added or subtracted, for example, y = cos x + 2 and
y = sin x + 2, the function will be moved up or down on the Y-axis, in
this case up 2.

• Following are graphs of *sine, cosine, tangent, secant, cosecant,* and
cotangent.

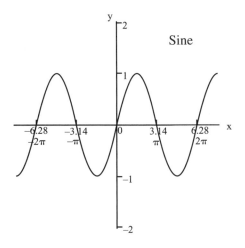

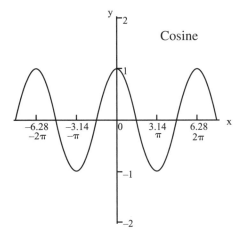

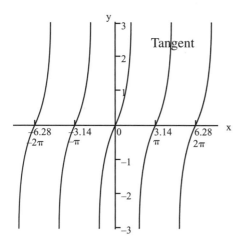

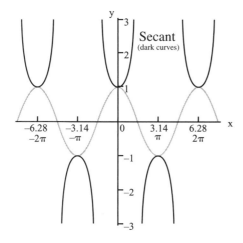

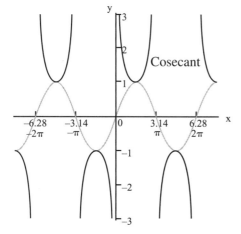

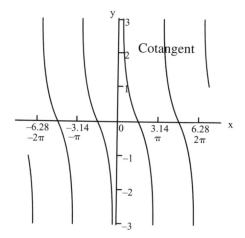

- *Inverse trigonometric functions* are *periodic* and their inverses are relations that are multivalued. Because of this, trigonometric functions are defined over a specific *interval* in their domain when their inverse is considered. A trigonometric function defined in this specific interval is often written with the first letter *capitalized*, e.g. *Sin, Cos, Tan.* For example, Sin x has a specific interval for its domain as $(-\pi/2 \le x \le \pi/2)$ and for its range as $(-1 \le y \le 1)$. (The range can be calculated in radians by taking the sine of $-\pi/2$, etc.)

The inverse of Sin is written Sin^{-1} or Arcsin and the relation between Sin and Sin^{-1} can be written:

$y = Sin\ x$ for $(-\pi/2 \le x \le +\pi/2, -1 \le y \le 1)$ and
$Sin^{-1}y = x$ or Arcsin $y = x$

For example, following are graphs for Arctangent and Arcsine:

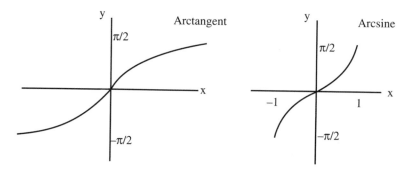

• The *domain and range values of trigonometric functions and their inverses* are provided below:

Function	Domain	Range
Sin x	$-\pi/2 \leq x \leq \pi/2$	$-1 \leq y \leq 1$
$Sin^{-1}x$	$-1 \leq x \leq 1$	$-\pi/2 \leq y \leq \pi/2$
Cos x	$0 \leq x \leq \pi$	$-1 \leq y \leq 1$
$Cos^{-1}x$	$-1 \leq x \leq 1$	$0 \leq y \leq \pi$
Tan x	$-\pi/2 < x < \pi/2$	$-\infty < y < \infty$
$Tan^{-1}x$	$-\infty < x < \infty$	$-\pi/2 < y < \pi/2$
Cot x	$0 < x < \pi$	$-\infty < y < \infty$
$Cot^{-1}x$	$-\infty < x < \infty$	$0 < y < \pi$
Sec x	$0 \leq x \leq \pi$, x not $\pi/2$	$-\infty < y \leq -1, 1 \leq y < \infty$
$Sec^{-1}x$	$-\infty < x \leq -1, 1 \leq x < \infty$	$0 \leq y \leq \pi$, y not $\pi/2$

Secant and Arcsecant are also defined as:

Sec x	$-\pi \leq x \leq 0$, x not $-\pi/2$	$-\infty < y \leq -1, 1 \leq y < \infty$
$Sec^{-1}x$	$-\infty < x \leq -1, 1 \leq x < \infty$	$-\pi \leq y \leq 0$, y not $-\pi/2$
Csc x	$-\pi/2 \leq x \leq \pi/2$, x not 0	$-\infty < y \leq -1, 1 \leq y < \infty$
$Csc^{-1}x$	$-\infty < x \leq -1, 1 \leq x < \infty$	$-\pi/2 \leq y \leq \pi/2$, y not 0

1.4. Circular Motion

• This section includes the principles of circular motion and harmonic motion.

• *Circular motion* and *harmonic motion* are important concepts in calculus and its applications. In circular motion, a point or particle moving in a circular path around the perimeter of a circle of radius 1 can be mapped using *sine* and *cosine*. The coordinates of the point or particle are given by (x = cos t, y = sin t), where t represents time. The coordinates for x and y on a unit circle satisfy $x^2 + y^2 = 1^2$. The speed of the particle is constant. One complete revolution of the particle around the circle corresponds to 2π radians.

If it takes one second for the particle to move around the circle, then it is moving at an *angular rate* of 1 revolution per second. Therefore, the particle moves around the circle with an *angular velocity* of 2π radians per second. The position of the particle is given by the angle ϕ, which is measured in radians. The rate of change of ϕ is the angular velocity of the particle. In other words, the angular velocity is the change in ϕ divided by the change in t, or $\omega = \Delta\phi/\Delta t$. If the motion is uniform, then $\phi = \omega t$. If $t = 1$, then $\phi = \omega$. If $\omega = 1$, then $\phi = t$.

• Because the coordinates of the particle traveling around the circle (at constant velocity) are given by $(\cos t, \sin t)$, as a particle moves around the circle, a point reflected onto the cosine axis that is following the movement of the particle will oscillate from side to side between $+1$ and -1, and a point reflected onto the sine axis that is following the movement of the particle will oscillate up and down between $+1$ and -1.

• Consider movement of a particle that begins at $t = 0$, $\phi = 0$, $x = \cos \phi = 1$ and $y = \sin \phi = 0$. As the particle moves upward and to the left toward the Y-axis, the particle reaches the top where $t = \pi/2 = \phi$, the upward velocity is $\cos \pi/2 = 0$ and the height is $\sin \pi/2 = 1$. Then the particle moves in the negative x direction and downward to $\phi = t = \pi$ where $x = \cos \pi = -1$ and $y = \sin \pi = 0$. The particle then moves down and to the right to $\phi = t = 3\pi/2$ where $\cos 3\pi/2 = 0$ and $\sin 3\pi/2 = -1$. Finally, the particle moves to the right and upward to $\phi = 2\pi = t = 0$ where $\cos 2\pi = 1$ and $\sin 2\pi = 0$.

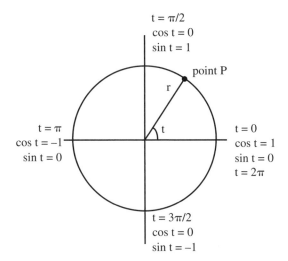

• As a point moves around a circle, at any given position the direction
of motion of the particle is tangent to the circle. For any *particle on the
circle* the following figure can be drawn. The *velocity tangent* to the circle
at point P has a cosine component cos t and a sine component sin t, and
points in the direction that the particle is moving. The acceleration this
particle experiences is *centripetal acceleration* which points inward along
the radius line.

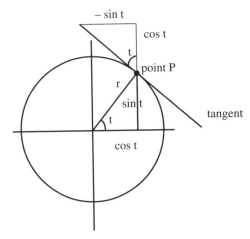

• In simple *harmonic motion*, or *oscillatory motion*, a particle or object
moves back and forth between two fixed positions in a straight line. The
connection between simple harmonic motion and uniform *circular motion*
can be visualized by projecting the image of a particle moving in a circular
path onto a screen (perpendicular to the plane of the circle). By projecting
the circular path from its side, the projected image looks like a particle
moving back and forth (or up and down) in a straight line.

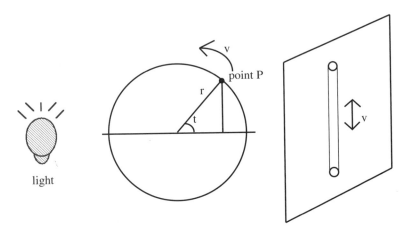

The shadow of the particle depicts the simple harmonic motion.

• Another means to visualize the connection between the particle moving around a circle in circular motion and its corresponding harmonic motion is to project the image of the particle as it moves around the circle onto the Y-axis.

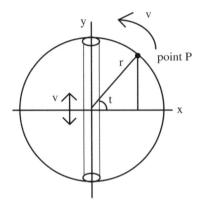

As the particle moves around the circle, the image oscillates up and down. The position of the particle

at the top is (cos $\pi/2$ = 0, sin $\pi/2$ = 1),
at the bottom is (cos $3\pi/2$ = 0, sin $3\pi/2$ = −1),
in the center on the way up is (cos 0 = 1, sin 0 = 0),
in the center on the way down is (cos π = −1, sin π = 0).

• By comparing the motion of this particle moving around the circle with its projection, it is evident that even though the velocity of the circular motion is constant, the velocity of the projection slows to a stop at each end (top and bottom). To evaluate the *velocity of the oscillatory motion*, it can be related to its *sine wave pattern*. The velocity is the rate of change of distance, and the slope of a curve at a given point represents the velocity at that point. By rotating a right triangle around a circle, the relationships between sine, cosine, distance, and velocity can be visualized. Also, these relationships can be visualized using the graphs of sine and cosine.

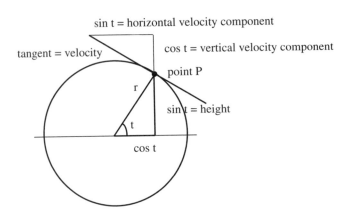

When the *distance* equals sin t, then the *velocity* equals cos t, and when the distance equals cos t, then the velocity equals −sin t.

• A *particle moving around the circle* can be compared to a *particle moving along the sine curve* to evaluate velocity. For example, the rate of change of position is zero at the top and bottom where t = π/2 and t = 3π/2 on the circle and on the sine curve. The *slope* of the sine curve at these points is zero. At these points the particle in straight line motion on the projected image in the circle is changing directions and comes to a stop as it turns where v = 0.

At t = 0, the particle is in the center of its projected strait line and is moving upward and has a corresponding slope on the sine curve of 1, therefore the velocity at this point is equal to 1. At t = π, the particle is in the center of its projected image and going straight down. At this position the corresponding slope on the *sine curve* is −1 and v = −1. The velocity of the particle at the center is at its greatest.

The slope at each point on the *sine curve* is given by the corresponding value of the *cosine curve* at that point.

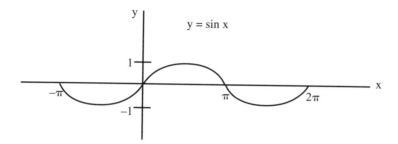

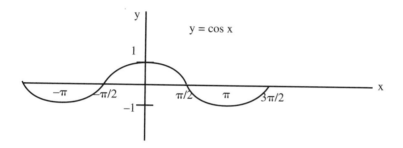

- *Cosine, sine* and *tangent* can be described by the equations:

$$\cos x = \cos(x + 2n\pi)$$
$$\cos x = \sin(\pi/2 + x)$$
$$\sin x = \sin(x + 2n\pi)$$
$$\tan x = \tan(x + n\pi)$$

where n is any integer and x is any real number.

1.5 Relationship Between Trigonometric and Exponential Functions

- *Trigonometric functions* and *exponential functions* are related to each other. Some important equations that define the relationship between these functions are listed in this section.

• Following are identities defining the relationships between trigonometric functions and exponential functions:

$$e^{ix} = \cos x + i \sin x$$

This is *Euler's identity* and defines the relationship between e^{ix}, cos x, and sin x.

$$e^{-ix} = \cos x - i \sin x$$
$$e^{i(-x)} = \cos(-x) + i \sin(-x)$$
$$\cos x = (1/2)(e^{ix} + e^{-ix})$$
$$\sin x = (1/2i)(e^{ix} - e^{-ix})$$
$$\cos \theta = (e^{i\theta} + e^{-i\theta})/2$$
$$\sin \theta = (e^{i\theta} - e^{-i\theta})/2i$$

Note that $i = \sqrt{-1}$. (See Section 1.10, "Complex Numbers," later in this chapter.) Also note that x or θ, etc., are used interchangeably for trigonometric and hyperbolic identities.

• The expansions for e^x, cos x, and sin x are:

$$e^x = 1 + x + x^2/2! + x^3/3! + x^4/4! + ... + x^n/n! + ...$$
$$\cos x = 1 - x^2/2! + x^4/4! - x^6/6! + ... + (-1)^{n-1}x^{2n-2}/(2n-2)! + ...$$
$$\sin x = x - x^3/3! + x^5/5! - x^7/7! + ... + (-1)^{n-1}x^{2n-1}/(2n-1)! + ...$$

1.6 Hyperbolic Functions

• Included in this section are equations for the hyperbolic functions: cosh, sinh, tanh, csch, sech, and coth.

• *Hyperbolic functions* are real, do not involve $i = \sqrt{-1}$ and are derived from the exponential functions e^x and e^{-x}.

The *hyperbolic cosine* is called *cosh* and is given by:

$$\cosh x = (1/2)e^x + (1/2)e^{-x}$$

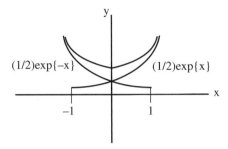

hyperbolic cosine, cosh

The *hyperbolic sine* is called *sinh* and is given by:

$$\sinh x = (1/2)e^x - (1/2)e^{-x}$$

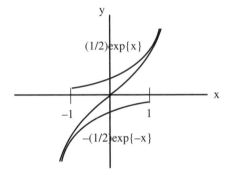

hyperbolic sine, sinh

Note that as x gets large, cosh x and sinh x approach $(1/2)e^x$.

• Like cosine, cosh is an *even function* such that:

cosh(–x) = cosh x and cosh 0 = 1

Like sine, sinh is an *odd function* such that:

sinh(–x) = –sinh x and sinh 0 = 0

• The properties that apply to cosh and sinh are similar to properties for cosine and sine; however, cosh x and sinh x do not involve the *i*. Examples of these properties include the following:

$(\cosh x)^2 - (\sinh x)^2 = 1$
$e^x = \cosh x + \sinh x$
$e^{-x} = \cosh x - \sinh x$
$\sinh^2 x = (1/2)(\cosh 2x - 1)$
$\cosh^2 x = (1/2)(\cosh 2x + 1)$
$\sinh(x \pm y) = \sinh x \cosh y \pm \cosh x \sinh y$
$\cosh(x \pm y) = \cosh x \cosh y \pm \sinh x \sinh y$

• Definitions for other hyperbolic functions include:

The *hyperbolic tangent*:
$\tanh x = (\sinh x / \cosh x) = (e^x - e^{-x})/(e^x + e^{-x})$

The *hyperbolic cosecant*:
$\operatorname{csch} x = 1 / \sinh x = 2/(e^x - e^{-x})$

The *hyperbolic secant*:
$\operatorname{sech} x = 1 / \cosh x = 2/(e^x + e^{-x})$

The *hyperbolic cotangent*:
$\coth x = \cosh x / \sinh x = (e^x + e^{-x})/(e^x - e^{-x})$

1.7 Polynomial Functions

• A *polynomial function*, in general, is a continuous function and its graph is a continuous curve. However, a polynomial function is not continuous if there are ratios of polynomial functions and the denominator is zero. At the point where a denominator is zero, the function is discontinuous.

• Remember polynomial expressions have the forms:

$3x^2 + 2x + 1$
$5x^4 + 3x^3 + 2x^2 - 8$

The *degree of a polynomial* is the highest value of an exponent in one of its terms.

• A polynomial function $P(x)$, can be written in the form

$P(x) = a_n x^n + a_{n-1} x^{n-1} + \ldots + a_1 x + a_0$

where the coefficients $a_1, a_0, a_n, a_{n-1}, \ldots$ are real numbers and $a_n \neq 0$.

1.8 Functions of More Than One Variable and Contour Diagrams

• This section includes functions that have more than one variable, and graphs and contour diagrams of these functions.

• Many functions depend on more than one variable. A *function that depends on two variables* can be written as

$z = f(x,y)$, where z is called the *dependent variable*, x and y are called the *independent variables*, and f represents the function.

• For example, the volume V of a pyramid depends on the height h and the area of its base A_b, which are independent variables. The function describing this is: $V = (1/3)hA_b$.

• Another example of a function that depends on more than one variable is the ideal gas law, $PV = nRT$, where P is pressure, V is volume, n is the number of moles in the sample, R is the universal gas constant (8.314 J/mol·k), and T is temperature. To study pressure $P = nRT/V$, vary one variable at a time while holding the others constant, or vary two at a time, etc. The data for functions that depend on more than one variable can be represented in tables or graphs in two or three dimensions. Temperature values for a system modeled by the ideal gas law can be listed on one axis and values for volume listed on another axis, such that resulting pressure values that correspond to a given temperature and volume will be within the table or on the third axis of a coordinate system.

• To graph a function $y = f(x)$ that depends on one variable, x values can be chosen and substituted into the function, and the corresponding y values can be calculated. Then the resulting points can be plotted on an XY coordinate system. The resulting graph represents the function and all of its points with coordinates (x,y), and is generally comprised of curves or lines.

The graph of a function $z = f(x,y)$ that depends on two variables represents the points with coordinates (x,y,z) and generally represents a *surface* in three-dimensional space. To graph $z = f(x,y)$, values for x and y can be chosen and z calculated. Values of x and y may represent data for a model system and be in a table format with x values listed down the side, y values listed across the top, and calculated z values inside the table.

• In general, *graphs of one-variable functions* form curves or straight lines, whereas *graphs of two-variable functions* form planes or surfaces represented in three-dimensional space (which comprises a family of *level curves* in the form of f(x,y) = constant). *Graphs of three-variable functions* form solids in four-dimensional space (which comprises a *family of level surfaces* in the form of f(x,y,z) = constant).

• A function having two independent variables f(x,y) can be represented by a surface, however this surface can be a member of a function having three independent variables F(x,y,z). Alternatively, one member of the family of surfaces in a three-variable function F(x,y,z) can be considered as representing the graph of a two-variable function f(x,y).

Therefore, a two-variable function represents a single surface and a three-variable function represents a family of level surfaces, where F(x,y,z) = f(x,y) – z for one of the surfaces and z = f(x,y) is the surface at F(x,y,z) = 0.

• Graphs of *linear one-variable functions* form straight lines with constant slopes. Graphs of *linear two-variable functions* form *planes* with the slopes of the columns being equal to each other and the slopes of the rows being equal to each other. Therefore, the slopes along lines in the plane parallel to the X-axis are the same, and the slopes of lines on the plane parallel to the Y-axis are the same.

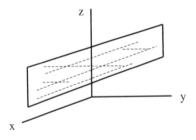

The *equation for a plane* represented by the linear function f(x,y) is:

$$f(x,y) = z = z_0 + m(x - x_0) + n(y - y_0)$$

where slopes in the x direction are designated by m, slopes in the y direction are designated by n, and the plane passes through point (x_0, y_0, z_0).

• When a *function has three or more independent variables*, it is more difficult to represent it in a graph, contour diagram, or table. Typically one or more variables can be held constant and the other(s) varied so that a table, graph, or contour diagram can be constructed.

• An example of a three-variable function is: $x^2 + y^2 + z^2 = Q$
Varying x, y, and z values will result in different Q values on the graphical representation which is a family of spheres nested inside of each other, with each one representing a different Q value.

• A *contour diagram* of curved surfaces can be depicted by connecting all of the points at the same height on the surface. The *level curves* form loops around the maximum point(s). As the height increases the loops get smaller. Level curves, or *contour lines*, are seen by slicing a surface with horizontal planes. The contour line at each height, h = z, is represented by f(x,y) = h. *Contour diagrams* representing planes contain parallel lines. A topographical map is an example of a contour diagram.

• **Example:** The graph of $f(x,y) = z = \sqrt{x^2 + y^2}$ is:

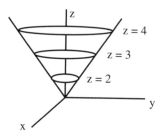

This graph has circular planes at each z value. When

f(x,y) = z = a constant, then a *contour map* or *diagram* can be drawn from the top-down perspective for

$f(x,y) = z = \sqrt{x^2 + y^2}$:

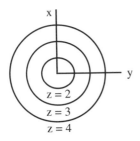

• **Example:** The graph of $f(x,y) = z = x^2 - y^2$ is:

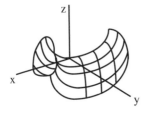

This graph forms a *saddle-shaped* surface. In this graph, there are two sets of parabolas, one set opening upward and the other set opening downward. Each curve corresponds to $f(x,y)$ when x is held constant and y is varied or when y is held constant and x is varied. Values for x, y, and z can be tabulated and plotted:

		\-2	\-1	0	1	2
		x values				
y	**2**	0	\-3	\-4	\-3	0
values	**1**	3	0	\-1	0	3
	0	4	1	0	1	4
	\-1	3	0	\-1	0	3

z values in grid

A contour diagram perspective of $f(x,y) = z = x^2 - y^2$ can also be depicted:

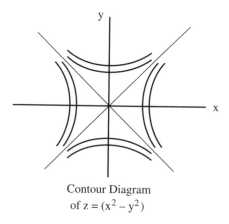

Contour Diagram
of $z = (x^2 - y^2)$

1.9 Coordinate Systems

• This section includes polar coordinates, cylindrical coordinates, and spherical coordinates.

• *Polar coordinates* describe points in a plane or in space, similar to rectangular *Cartesian coordinates*. The difference is that in polar coordinates, there is an r-coordinate that maps the distance of a point from the origin of the coordinate system, and there is a θ-coordinate that measures the angle the r-ray makes from the horizontal positive X-axis.

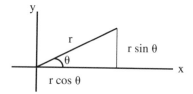

The relationship between polar and *rectangular coordinates* can be visualized in the figure using the *Pythagorean Theorem* for a right triangle $r^2 = x^2 + y^2$. The r-coordinate is the *hypotenuse* and measures the distance from origin to the point of interest. The angle θ between r and the positive part of the X-axis can be described by $\tan \theta = y/x$. The relationships between these two coordinate systems are:

$$r = \sqrt{x^2 + y^2}$$

$\tan \theta = y/x$ or $\theta = \tan^{-1}(y/x)$

$x = r \cos \theta$

$y = r \sin \theta$

• In three dimensions, polar coordinates become *cylindrical coordinates* and are given in terms of r, θ, and z, where:

$$x = r \cos \theta$$
$$y = r \sin \theta$$
$$z = z$$
$$r = \sqrt{x^2 + y^2}$$

When comparing the Cartesian and cylindrical coordinate systems, the x- and y-components of the Cartesian coordinate system are expressed in terms of polar coordinates, and the z-component is the same component as in the Cartesian system. The r-component is measured from the Z-axis, the θ-component measures the distance around the Z-axis, and the z-component measures along the Z-axis.

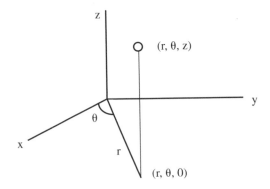

• Following are examples using cylindrical coordinates:

(a.) To find x and y if r and θ are given, for example:

r = 5 and θ = π/2, simply calculate
x = r cos θ and y = r sin θ:
x = (5) cos(π/2) and y = (5) sin(π/2).

Alternatively, to find r and θ if x and y are given, for example, x = 2 and y = 3, calculate

$$r = \sqrt{x^2 + y^2} \text{ and } \theta = \tan^{-1}(y/x):$$

$$r = \sqrt{2^2 + 3^2} \text{ and } \theta = \tan^{-1}(3/2)$$

(b.) A *circle on a coordinate system* can be represented by the equations
r = cos θ or r = sin θ, where substituting values of θ around the
coordinate system will produce points on the circle.

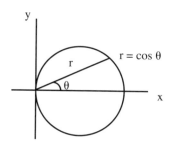 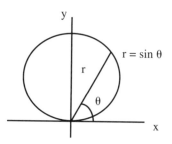

(c.) If r = θ, a *spiral of Archimedes* can be plotted:

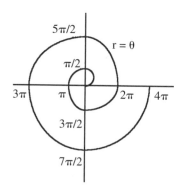

(d.) A *triangular wedge section* can be represented in cylindrical coordinates by considering that the section ($\Delta\theta/2\pi$) is a part of a whole area
of the circle πr^2 that projects along the Z-axis to create a volume.

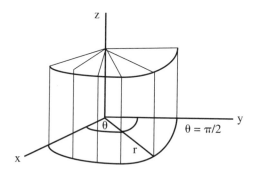

• Another coordinate system that is related to Cartesian coordinates is *spherical coordinates*. In three dimensions, spherical coordinates are expressed in terms of ρ, θ, and ϕ, where ρ can range from 0 to ∞, θ can range from 0 to 2π, and ϕ can range from 0 to π. In spherical coordinates, the ρ component is measured from the origin, the θ component measures the distance around the Z-axis, and the ϕ component measures down from the Z-axis and is referred to as the *polar angle*. Note that ρ is measured from the origin rather than the Z-axis as is the case with r in cylindrical coordinates. Also, θ and ϕ are similar to longitude and latitude on a globe. Spherical coordinates can be defined in terms of Cartesian coordinates, x, y, and z as:

$$x = \rho \cos \theta \sin \phi$$
$$y = \rho \sin \theta \sin \phi$$
$$z = \rho \cos \phi$$
$$\rho = \sqrt{x^2 + y^2 + z^2}$$

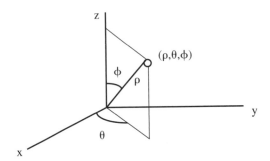

1.10 Complex Numbers

• This section includes complex numbers; imaginary numbers; adding, subtracting, multiplying, and dividing complex numbers; the complex plane; and expansions involving *i*.

• *Complex numbers* are numbers involving $\sqrt{-1}$. Because there is no number that when squared equals −1, the symbol *i* was introduced, such that $\sqrt{-x} = i\sqrt{x}$, where x is a positive number and $(i)^2 = -1$.

• Complex numbers involve i and are generally in the form $(x + iy)$, where x and y are real numbers. In this expression, the x term is referred to as the real part and the iy term is referred to as the imaginary part. A real number multiplied by i forms an *imaginary number*, such that:

real number $\times i$ = imaginary number.

A real number added to an imaginary number forms a complex number, such that:

real number + (real number)(i) = complex number.

• *Complex numbers are added or subtracted* by adding or subtracting the real terms and imaginary terms separately. The result is in the form $(x + iy)$. For example:

$(1 + 2i) + (3 + 4i) = 1 + 3 + 2i + 4i = 4 + 6i$

• *Complex numbers are multiplied* as ordinary binomials, and $(i)^2$ is replaced by -1. For example:

$(1 + 2i)(3 + 4i) = (1)(3) + (1)(4i) + (2i)(3) + (2i)(4i)$
$= 3 + 4i + 6i + 8(i)^2 = 3 - 8 + 10i = -5 + 10i$

• *Complex numbers are divided* by first multiplying the numerator and denominator by what is called the *complex conjugate* of the denominator, then the numerator and denominator are divided and combined as with multiplication. For example, the complex conjugate of $(3 + 2i)$ is $(3 - 2i)$. The product of a complex number and its conjugate is a real number. Remember to replace $(i)^2$ by -1 during calculations. For example, divide the following:

$(1 + 2i) \div (3 + 4i) = (1 + 2i)(3 - 4i) \div (3 + 4i)(3 - 4i)$
$= (3 - 4i + 6i - 8i^2) \div (9 - 12i + 12i - 16i^2)$
$= (3 + 2i - 8(-1)) \div (9 - 16(-1)) = (11 + 2i)/25 = 11/25 + 2i/25$

(*See Master Math: Basic Math and Pre-Algebra*, Section 1.17 for examples of the above principles.)

• Complex numbers can correspond to points in a coordinate system, sometimes called the *complex plane*. For example, $3 + 4i$ corresponds to $x = 3$ and $y = 4$, where the X-axis is real and the Y-axis is imaginary. Using polar coordinates $x = r \cos \theta$ and $y = r \sin \theta$, the complex plane is described using:

$$x + iy = r \cos \theta + i\, r \sin \theta = r(\cos \theta + i \sin \theta)$$

where $\cos \theta + i \sin \theta = e^{i\theta}$ is Euler's formula.

imaginary-axis

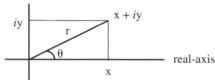

• *Expansions of e and trigonometric functions* involve imaginary numbers. For example:

$$\cos \theta = 1 - \theta^2/2! + \theta^4/4! - \ldots$$
$$i \sin \theta = i\theta - i\theta^3/3! + i\theta^5/5! - \ldots$$
$$e^{i\theta} = 1 + i\theta - \theta^2/2! - i\theta^3/3! + \theta^4/4! + \ldots$$

1.11 Parabolas, Circles, Ellipses, and Hyperbolas

• This section includes parabolas, circles, ellipses, and hyperbolas, and their equations and definitions.

• Parabolas, circles, ellipses, and hyperbolas are important curves in calculus. They are often referred to as *conic sections*, because each of these curves can be represented as the intersection of a plane with right circular cones.

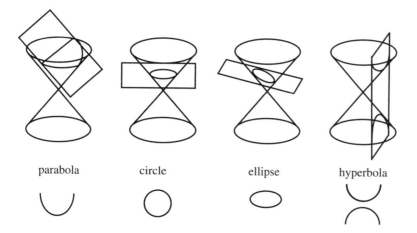

Examples of applications where conic sections are used for modeling include ellipses for orbits of planets, parabolas for the path of a projectile, and hyperbolas for the reflection of sound.

Parabolas

- The equations for a *parabola* are:

  $y = ax^2 + bx + c$ with a vertical axis
  $x = ay^2 + by + c$ with a horizontal axis

- Following are vertical-axis parabolas:

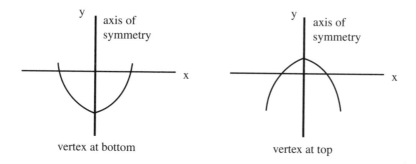

- In the vertical form of the equation, $y = ax^2 + bx + c$, if a is positive, the parabola is open at the top with the *vertex* at the bottom. Conversely, if a is negative, the parabola is open at the bottom with the vertex at the top. For $y = ax^2 + bx + c$, the graph crosses the X-axis at $y = 0$, and the *vertex point* is a minimum or maximum point (where $dy/dx = 0$).

• For a vertical-axis parabola, if the parabola lies above its vertex, then the y-coordinate (x, y) of its vertex is the smallest y value of the parabola that satisfies the equation for that parabola. Conversely, if the parabola lies below its vertex, then the y-coordinate (x, y) of its vertex is the largest y value of the parabola that satisfies the equation for that parabola.

• The *axis of symmetry* can be drawn through the center of a parabola to divide it in half. The equation for the axis of symmetry in a vertical parabola is $x_v = -b/2a$. For example, if the solution to this equation is $x_v = 2$, then a vertical line through the point 2 on the X-axis can be drawn to represent the axis of symmetry.

• The *vertex point of a parabola* can be found by substituting x_v into the equation, $y = ax^2 + bx + c$, (if x_v is known) and solving for the corresponding y or y_v value, resulting in (x_v, y_v). The vertex point of a parabola with a vertical axis can be found using the equation for the parabola as follows: (a.) separating the terms that contain x from the terms not containing x, (b.) completing the square on the terms that contain x, (c.) setting each side of the equation equal to zero, and (d.) solving the resulting equations for x and y resulting in (x_v, y_v).

For example, if $y = x^2 - 2x - 3$, what is the vertex point?

Rearrange:

$x^2 - 2x = y + 3$

Complete the square by finding 1/2 of the coefficient b (b = 2). Square 1/2 of the coefficient b, $(b/2)^2$, and add the result to each side of the equation:

$(b/2)^2 = (-2/2)^2 = (-1)^2 = 1$
$x^2 - 2x + 1 = y + 3 + 1$
$x^2 - 2x + 1 = y + 4$

Factor the resulting perfect square, and set each side of the equation equal to zero and solve:

$(x - 1)(x - 1) = y + 4$

$(x - 1)^2 = 0$

$x = 1$

$y + 4 = 0$

$y = -4$

Therefore, the vertex point is (1,–4).

• To *graph a parabola*, which is a *quadratic equation*, one method involves finding the x component of the vertex point, $x_v = -b/2a$, and substituting x_v into the equation, $y = ax^2 + bx + c$, and solving for the corresponding y or y_v value, resulting in (x_v, y_v). Then, choose other values for x on both sides of x_v and solve for their corresponding y values using the original equation. Finally, the points can be plotted and the parabola sketched.

• It is possible to *solve a quadratic equation graphically* as follows:

(a.) Write the equation in the standard form $y = ax^2 + bx + c$.

(b.) Graph the parabola by identifying $x_v = -b/2a$, substitute x_v into the equation $y = ax^2 + bx + c$ and solve for y_v resulting in the vertex point (x_v, y_v). Then choose other values for x on both sides of x_v and solve for their corresponding y values using the original equation to plot the parabola.

(c.) Determine the solutions for x (called the roots of the equation because of the x^2 term) by estimating the two points where the parabola crosses the X-axis (at $y = 0$).

• The *focus of a parabola* is a point on the axis of symmetry where any ray (e.g. light, sound, etc.) coming toward the bottom of the parabola, parallel with the axis of symmetry is reflected to.

An example of the use of the *focus point* is a receiver of radio waves or TV signals where the rays are concentrated at the focus. This principle applies in reverse as well. When light energy is emitted from a focus point and reflected off the inner surface of the parabola, it will point out of the parabola parallel to the axis of symmetry.

• The *directrix* is a line that exists perpendicular to the axis of symmetry such that every point on the parabola is the same distance from the focus point as it is from the directrix line. Therefore, the distance from the vertex to the focus (d2 on graph) is equal but opposite to the distance from the vertex to the directrix (d1 on graph).

• The following is a *horizontal-axis parabola* with its vertex at the left:

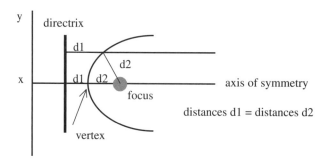

Ellipses and Circles

• Equations for *circles* are written in the forms:

$$x^2 + y^2 = r^2$$

where r is the radius and r > 0, and the circle is located at the origin of a coordinate system.

For a circle centered at a point (x = p, y = q) other than the origin, the equation becomes:

$$(x - p)^2 + (y - q)^2 = r^2$$

• The equations for *ellipses* are written in the forms:

$$(x^2/a^2) + (y^2/b^2) = 1$$

where a ≠ b, a > 0, b > 0, origin at (0,0).

If a = b, the ellipse becomes a circle. The equation for an ellipse with a = b = r is the equation for a circle:

$$x^2 + y^2 = r^2$$

For an ellipse centered at a point (x = p, y = q) other than the origin, the equation becomes:

$$((x - p)^2/a^2) + ((y - q)^2/b^2) = 1$$

where a ≠ b, a > 0, b > 0, origin at (p,q).

• The *equations for ellipses and circles* having the form $(x^2/a^2) + (y^2/b^2) = 1$, with origin at $(0,0)$ and $a \neq b$ for ellipses and $a = b$ for circles, can be solved for y as follows:

$$y/b = \pm[1 - (x^2/a^2)]^{1/2} = \pm[(a^2/a^2) - (x^2/a^2)]^{1/2}$$
$$= \pm[(a^2 - x^2)/a^2)]^{1/2} = \pm (1/a)[a^2 - x^2]^{1/2}$$

Therefore, $y = \pm (b/a)[a^2 - x^2]^{1/2}$

where the $(+)$ values represent the top half of an ellipse or circle and the $(-)$ values represent the bottom half of an ellipse or circle. The curve crosses from $(+)$ to $(-)$ or $(-)$ to $(+)$ at $y = 0$, $x = a$ and $y = 0$, $x = -a$, respectively. The maximum and minimum of the curves are at $y = b$ and $y = -b$.

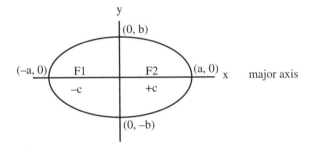

• A circle has one *focus* at the center. An ellipse has two *foci*, designated F1 and F2, along the *major axis* on either side of the center. The sum of the distances of the foci to all the points on an ellipse is 2a. Therefore at any point on the ellipse: (the distance to F1) + (the distance to F2) = 2a.

The ellipse can also be described by:

$$[(x - c)^2 + y^2]^{1/2} + [(x + c)^2 + y^2]^{1/2} = 2a$$
$$= \text{distance from F1} + \text{F2 to (x,y)},$$

where $+c$ and $-c$ represent the location of F1 and F2 on the major axis.

• Because the distance from F1 to F2 is 2a, as sound waves, etc., are reflected off the ellipse, a sound generated at F1 will be concentrated at F2. (Note that there is also a direct path between F1 and F2 where sound is not reflected.)

• To plot an equation of an ellipse, choose values for x and solve for the corresponding y values. Alternatively, set x = 0 and solve for the corresponding y value and set y = 0 and solve for the corresponding x value.

• To plot an equation of a circle, choose values for x and solve for the corresponding y values. Alternatively, set x = 0 and solve for the corresponding y value and set y = 0 and solve for the corresponding x value. If a circle has its origin at (0,0) it is possible to choose x values in one quadrant of the coordinate system and use symmetry to complete the circle.

Hyperbolas

• The *equations for hyperbolas* can be written in the forms:

$$(x^2/a^2) - (y^2/b^2) = 1 \quad \text{or} \quad -(x^2/a^2) + (y^2/b^2) = 1$$

where a and b have opposite signs and the origin is at (0,0).

For a *hyperbola* located at a point (x = p, y = q) other than the origin, the equation becomes:

$$((x - p)^2/a^2) - ((y - q)^2/b^2) = 1$$

• The equation for a hyperbola can be solved for y:

$$y/b = \pm[1 + (x^2/a^2)]^{1/2} = \pm[(a^2/a^2) + (x^2/a^2)]^{1/2}$$
$$= \pm[(a^2 + x^2)/a^2)]^{1/2} = \pm(1/a)[a^2 + x^2]^{1/2}$$

Therefore, $y = \pm(b/a)[a^2 + x^2]^{1/2}$

where the (+) expression represents the side of the hyperbola where $y \geq b$, and the (−) expression represents the side of the hyperbola where $y < b$.

In the following figure, v1 and v2 are *vertexes* at (0,b) and (0,−b), and F1 and F2 are *foci*. The ray drawn coming toward one focus, F2, and contacting the outside of that side of the hyperbola will be reflected to the other focus, F1.

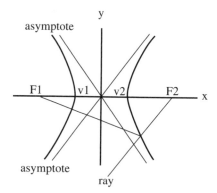

- A hyperbola can be drawn along the X-axis or Y-axis and is symmetric with respect to its axis. The diagonal lines are called *asymptotes*, and each hyperbola has two asymptotes such that the curve of a hyperbola approaches its asymptotes. The *equations for the asymptotes* have zero replaced for the constant terms, and for a hyperbola centered at (0,0) in a coordinate system, the equations for the asymptotes are $y = \pm(b/a)x$, and the slopes are $+b/a$ and $-b/a$.

- The foci of the hyperbola are inside the curve of each side such that for the points on the hyperbola, the difference between the distances to the foci is 2b (see preceding figure).

- Another equation for a hyperbola is $xy = k$. If k is positive, the hyperbola will graph in the upper right and lower left quadrants. Conversely, if k is negative, the hyperbola will graph in the upper left and lower right quadrants.

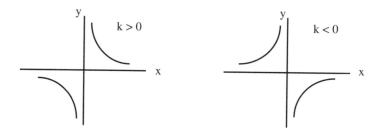

• To plot an equation of a hyperbola, choose values for x and solve for the corresponding y values. Because the graph of a hyperbola is symmetric with respect to both axes, the points plotted in one quadrant (e.g. $x \geq 0$, $y \geq 0$) will mirror the points in the opposite quadrant.

Chapter

2

The Derivative

2.1 The Limit

• This section includes a summary of the limit, its applications, conditions for estimating an infinite series, and convergence.

• The *limit* is used in the estimation of infinite sums and in the definitions of the derivative and the integral. In general, the limit is used to describe *closeness* of a function to a value when the exact value cannot be identified. The limit is also used to determine if functions are continuous or discontinuous. Whether the graph of a given function is a smooth and continuous curve or line, or whether there are breaks or holes present can be determined using the limit.

• If the $\lim_{x \to a} f(x) = L$, then it is said that the function $f(x)$ gets close to and may equal some number L as x approaches and gets close to some number a.

• Consider a simple example where the limit as x approaches 1 is taken of function $f(x) = x^2 + 2$:

$$\lim_{x \to 1} f(x) = \lim_{x \to 1} (x^2 + 2)$$

By substituting numbers for x that get closer to the number 1, it can be shown that as x gets closer to 1, or $\lim_{x \to 1}$, then x^2 gets closer to 1 and $f(x)$ gets closer to 3.

• In more complicated examples, simply substituting the desired number into the function is not possible. For example, consider the limit:

$$\lim_{x \to 2} [(x^2 - 4) / (x - 2)]$$

If x = 2 is substituted into this equation, the result is 0/0, which is undefined. In this equation, factoring can be used to determine if the limit exists. (See Section 5.7 in *Master Math: Algebra* for a review of factoring polynomials.)

The factoring of $[(x^2 - 4) / (x - 2)]$ results in: x + 2.
Taking the limit results in: $\lim_{x \to 2}(x + 2) = 4$.
Therefore, as x gets close to 2, then x + 2 gets close to 4.

• There are many cases where the limit of a function does not exist. For example, consider the limit as y gets close to zero for the function $f(y) = 1/y$: $\lim_{y \to 0}(1/y)$.

Because 1/0 is undefined, it is possible to substitute values for y approaching zero:

If y = 0.1, f(y) = 1/0.1 = 10
If y = 0.01, f(y) = 1/0.01 = 100
If y = 0.001, f(y) = 1/0.001 = 1,000
If y = − 0.01, f(y) = 1/−0.01 = −100
If y = −0.001, f(y) = 1/−0.001 = −1,000

As y gets close to 0, 1/y does *not* get closer to any number. The magnitude of 1/y actually increases. Therefore, $\lim_{y \to 0}(1/y)$ approaches ∞, and the limit does not exist.

• If the *limits of the functions* f(x) and g(x) exist, then the rules governing limits of functions that are *added, subtracted, multiplied, divided,* and *raised to a power* are:

$$\lim_{x \to a}(f(x) + g(x)) = \lim_{x \to a}f(x) + \lim_{x \to a}g(x)$$
$$\lim_{x \to a}(f(x) - g(x)) = \lim_{x \to a}f(x) - \lim_{x \to a}g(x)$$
$$\lim_{x \to a}(f(x) \times g(x)) = \lim_{x \to a}f(x) \times \lim_{x \to a}g(x)$$
$$\lim_{x \to a}(f(x) \div g(x)) = \lim_{x \to a}f(x) \div \lim_{x \to a}g(x), \text{ if } \lim_{x \to a}g(x) \neq 0$$
$$\lim_{x \to a}(f(x))^y = (\lim_{x \to a}f(x))^y$$

- The *limit can be used to estimate the sum of an infinite series.* If the progression or sequence is infinite and therefore there is an infinite number of terms, then the sum cannot be calculated. However, under certain conditions the sum of an *infinite series* can be estimated.

- Conditions that determine if the *sum of an infinite series can be estimated* include the following.

(a.) If an infinite series has a limit, it is said to *converge* and the sum can be estimated. In other words, as the terms in an infinite series are added, beginning with the first term, if with each additional term added the sum approaches some number, then the series has a limit and converges and the sum can be estimated.

(b.) A condition for *convergence* for the infinite series:

$$\sum_{n=1} a_n$$

is that a_n must approach zero as n approaches infinity. Although this condition must occur for a series to converge, there are cases where this condition is true but the series still diverges.

(c.) If an infinite series has no limit, it is said to *diverge* and the sum cannot be estimated. In other words, if instead as each additional term is added the sum approaches infinity, then the series has no limit and diverges and the sum cannot be estimated.

- To *estimate an infinite series*, it must be determined whether the series has a limit and converges and what happens to the sum as the number of terms approach infinity. For an infinite series describing the sum of a_n from n = 1 to n = ∞:

$$\sum_{n=1} a_n$$

If this series has a limit and converges to L it becomes:

$$\lim_{n \to \infty} \sum_{n=1} a_n = L$$

For convergence to zero, the values of a_n must become small at some small number and remain below that number. The small number may be on the order of 10^{-10}, etc. The generally accepted notation for representing this small number that a converging sequence will approach and remain below is ε (Epsilon). The general rule is, for any value of ε, there is a number N where $|a_n| < ε$ if $n > N$.

For convergence to numbers other than zero, the general rule is, for any ε, there is a number N where $|a_n - L| < ε$ if $n > N$. In this case, $\lim_{n \to \infty} a_n = L$. Convergence occurs when values for a_n approach L and remain in a range of $L + ε$ and $L - ε$.

• When two sequences, a_n and b_n, each converge such that $\lim_{n \to \infty} a_n = L$ and $\lim_{n \to \infty} b_n = L^*$, then the following are true:

$\lim_{n \to \infty}(a_n + b_n) = L + L^*$

$\lim_{n \to \infty}(a_n - b_n) = L - L^*$

$\lim_{n \to \infty}(a_n\, b_n) = L\, L^*$

$\lim_{n \to \infty}(a_n/b_n) = L/L^*$ provided $L^* \neq 0$

2.2 Continuity

• This section provides a brief summary of continuity including the definition of a continuous function, a continuable function, examples of continuous functions, discontinuous functions, conditions of discontinuity, and visualizing continuity.

• A function is considered *continuous* at $x = a$, if $\lim_{x \to a}$ exists, such that the $\lim_{x \to a} f(x) = f(a)$ and $f(a)$ is defined. If $f(x)$ is continuous at $x = a$, then $f(x) \to f(a)$ as $x \to a$. For a function to be defined as a *continuous function*, it must be continuous at every point where it is defined.

• A function is called *continuable* if the definition of continuous can be applied to all x values such that the function is continuous at all x values. For example, $f(x) = 1/x$ is not continuable but is continuous by definition because it is "not defined" at $1/x$ where $x = 0$. A function is continuous at x if as $\Delta x \to 0$, then $[f(x + \Delta x) - f(x)] \to 0$, where Δx represents some small change or increment in x.

• A *polynomial function* is an example of a continuous function that is continuous everywhere and its graph is a continuous curve. (With the exception of ratios of polynomial functions with zero denominators.)

Other continuous functions include *exponential, sine, cosine,* and *rational functions* on intervals where their denominators are not zero.

• A function that is not continuous may be *discontinuous* at a single point. For example, the function $f(x) = 1/x$ is continuous except at $x = 0$ where $1/0$ is undefined. Therefore, $\lim_{x \to a} f(x) = \lim_{x \to a} 1/x = 1/a$. In general, the *graph of a continuous function* has no holes or breaks.

• The following graphs are of (a.) a function that is *discontinuous at a point,* (b.) a function that is discontinuous at more than one point and has a *jump,* (c.) function $f(x) = 1/x^2$, and (d.) function $f(x) = 1/x$:

(a.)
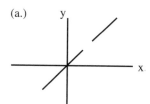

(b.)
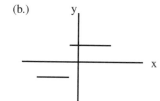

(c.)
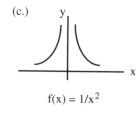

$f(x) = 1/x^2$

(d.)
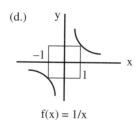

$f(x) = 1/x$

• The following are conditions where *discontinuity* exists:

(a.) If $\lim_{x \to a} f(x)$ exists and is equal to L, but $f(x)$ does not exist when $x = a$ so that $f(a) \neq L$, then the graph of $f(x)$ is discontinuous at the point $x = a$. In this case there is only discontinuity at a single point.

(b.) If $\lim_{x \to a} f(x)$ does not exist because as x approaches a from either $x > a$ or $x < a$, the value of $f(x)$ approached from $x > a$ is different from the value $f(x)$ approached from $x < a$, then the graph of $f(x)$ is discontinuous and has a *jump* at point a $= x$.

(c.) If $\lim_{x \to a} f(x)$ does not exist because as x approaches a, the absolute value $|f(x)|$ gets larger and larger then the graph is "infinitely discontinuous" at $x = a$.

• A means to visualize whether a function is *continuous* involves use of symbols such as Epsilon ε and Delta δ to define regions in question on the X and Y axes of the graph of a function. Consider the limit of the function f(x) where $\lim_{x \to a} f(x) = L$, and has the following properties:

(a.) The limit exists.

(b.) ε represents an error tolerance allowed for L.

(c.) δ represents the distance that x is from x = a.

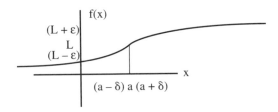

In this graph a value for ε can be chosen and δ(ε) results. Conversely, a value for δ can be chosen and ε(δ) results.

As $\lim_{x \to a} f(x) = f(a) = L$, providing the limit exists, where:

$$L + \varepsilon = f(a) + \varepsilon, \text{ or } L = f(a), \text{ also } L - \varepsilon = f(a) - \varepsilon,$$

then the following is true:

$$f(a) + \varepsilon > f(x) > f(a) - \varepsilon, \text{ or equivalently, } |f(x) - L| < \varepsilon.$$

For every chosen number ε where ε > 0, there is a positive number for δ(ε) that results.

For a chosen f(x), x must be within (a − δ) and (a + δ) such that $a - \delta < x < a + \delta$, or equivalently, $0 < |x - a| < \delta$.

Therefore, as x gets close enough to a, then $|f(x) - L| < \varepsilon$, and the closeness of x is defined to have a tolerance of δ such that when $0 < |x - a| < \delta$ then $|f(x) - L| < \varepsilon$.

2.3 Differentiability

• This section summarizes the concept of differentiability. (See the next section for the definition of the derivative.)

• A function is *differentiable* at any point where it has a derivative. A function that has a derivative at every point is differentiable everywhere. At any point where f(x) has a derivative, the function is *continuous*. There can be a point where f(x) is continuous but no derivative exists, such as where the graph turns a corner without a hole or jump.

• A graph of a function has a *derivative* and is therefore differentiable at a point if a *tangent line* can be drawn at that point.

• If for a given point on the graph of a function the derivative does not exist, then that point may be (a.) at the end of the curve of the function; (b.) at a corner on the curve; (c.) at a location where the tangent line is a vertical line and therefore has no slope; or (d.) at a location on the graph that is discontinuous such as if one point is missing or there is a jump in the curve of the function.

2.4 The Definition of the Derivative and Rate of Change

• This section includes the definition of the derivative, notation, developing the definition of the derivative, calculating velocity using the derivative, the average rate of change, and the instantaneous rate of change.

• The *derivative* is used to describe the rate of change of something such as *velocity*, as well as the concept of the *tangent* to a curve. Applications of the derivative include tangents, slopes, rates of change, curvilinear and straightline motion, maxima, minima, and tests for extrema.

• *Notation for the derivative* of a function f(x) includes:

$$\frac{df(x)}{dx}, \ \frac{d}{dx} f(x), \ f'(x), \ \mathbf{D}f(x), \ \mathbf{D_x}f(x)$$

If $y = f(x)$ then the derivative $f'(x)$ can be written (dy/dx).

Notation for taking *second* derivatives includes:

$$\frac{d^2 f(x)}{dx^2}, \ \frac{d^2}{dx^2} f(x), \ f^2(x), \ f''(x), \ \mathbf{D}^2 f(x), \ \mathbf{D_x}^2 f(x)$$

Notation for the *nth* derivative includes:

$$\frac{d^n f(x)}{dx^n}, \frac{d^n}{dx^n} f(x), f^n(x), \mathbf{D}^n f(x), \mathbf{D}x^n f(x)$$

• The *time rate of change* of an object in motion such as a car, plane, pitcher's fast ball, etc., is the rate of change of distance with respect to time and is called *velocity*. The velocity is the derivative or equivalently the rate of change of distance with respect to time. Velocity can be positive or negative with respect to a reference point, but *speed* is the magnitude of velocity and is always positive or zero.

• To consider rate of change, remember that distance equals rate times time, d = rt, therefore, rate = (distance / time).

The time rate of change of distance is velocity and *average velocity* = (change in distance / change in time).

Also, the time rate of change of velocity is *acceleration* and acceleration = (change in velocity / change in time).

• To develop the *definition of the derivative*, consider the velocity of an airplane flying from the east coast to the west coast.

The distance the airplane is from its starting point or any defined reference point is a function of time (depends on time) or f(t). (In this example, f is the distance function.)

At time = t, the airplane is f(t) units from the starting or reference point. (The units could be hours.)

At time = t + h, the airplane is f(t+h) units from the starting or reference point and *h represents an increment of time.*

The change in the position of the airplane during the increment of time h is f(t+h) − f(t).

The rate of change of the distance with respect to time between time = t and time = t+h is the average velocity of the airplane. The *average velocity* during this time period is:

$$\text{average velocity} = \frac{f(t+h) - f(t)}{h}$$

To find the velocity of the airplane at a particular point when time = t, shrink the time increment h surrounding time t. The velocity at the point where time = t is called the *instantaneous velocity*, and is determined by taking the limit as the increment of time h shrinks to zero:

$$\text{velocity at time } t = v(t) = \lim_{h \to 0} \frac{f(t+h) - f(t)}{h}$$

As h gets close to zero (but not equal to zero), the time increment h and the distance $f(t+h) - f(t)$ will get smaller.

Because velocity is the derivative of distance, then the definition of the derivative with respect to time of the distance function f(t) can be written:

$$\frac{df(t)}{dt} = \lim_{h \to 0} \frac{f(t+h) - f(t)}{h}, \text{ provided the limit exists.}$$

• The *velocity* at time t or v(t) can be determined using the *definition of the derivative* and the following procedure:

(a.) Determine $f(t+h)$ and $f(t)$.

(b.) Subtract $f(t)$ from $f(t+h)$.

(c.) Divide $[f(t+h) - f(t)]$ by h.

(d.) Take the limit as h approaches zero.

• **Example:** Find the velocity at t = 2 hours, if the distance in miles is represented by $f(t) = 3t^2$.

$f(t+h) = 3(2 + h)^2 = 3(4 + 4h + h^2) = 12 + 12h + 3h^2$
$f(t) = 3(2)^2 = 12$
$f(t+h) - f(t) = (12 + 12h + 3h^2) - 12 = 12h + 3h^2$
$(f(t+h) - f(t)) \div h = (12h + 3h^2) \div h = (12h)/h + (3h^2)/h = 12 + 3h$
$\lim_{h \to 0}(12 + 3h) = 12 = $ the velocity at 2 hours, v(2hrs)

Therefore, the velocity at t = 2 hours is 12 miles/hour.

• The *definition of the derivative* with respect to x, rather than with respect to time, can be written:

$$\frac{df(x)}{dx} = \lim_{h \to 0} \frac{f(x+h) - f(x)}{h}$$

• It is possible to use the definition of the derivative to determine the average rate of change or the instantaneous rate of change of a function. In general, the rate of change represents how fast or slow a function changes from one end of the interval to the other end, relative to the size of the interval (given by h).

• The *average rate of change* of f over an interval from some value of x to some value of x + h is given by:

$$\frac{f(x+h)-f(x)}{h}$$

The average rate of change is equivalent to the *slope* of a line drawn between two points on the graph of a function f(x) represented by x values between the value of x = a to the value of x = (a + h).

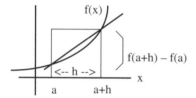

• The *instantaneous rate of change* of f(x) at some point x is given by the following expression, which represents the average rate of change over smaller and smaller intervals. This expression defines the derivative of f at some point x.

$$\lim_{h \to 0} \frac{f(x+h)-f(x)}{h}$$

The instantaneous rate of change is equal to the slope of the graph of the function at some *point* on the curve, or equivalently the instantaneous rate of change is equal to the slope of a line drawn tangent to the curve at that point.

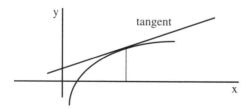

2.5 Δ (delta) Notation and the Definition of the Derivative

• This section introduces *delta* Δ notation for the definition of the derivative.

• An alternative notation for writing the definition of the derivative is to use Δx in place of h and Δy (or Δf) in place of f(x + Δx) − f(x). Using this Δ *(delta) notation* the derivative with respect to x can be written:

$$f'(x) = \lim_{\Delta x \to 0} \frac{f(x + \Delta x) - f(x)}{\Delta x} = \lim_{\Delta x \to 0} \frac{\Delta y}{\Delta x} = \frac{dy}{dx}$$

In the context of an XY coordinate system, $\Delta y/\Delta x$ represents the average rate of change of y per unit change in x over the interval of a curve of a function between x and $x + \Delta x$. Similarly, dy/dx represents the instantaneous rate of change in y per unit change in x at some point (x,f(x)).

This is also sometimes written:

$$\frac{\Delta y}{\Delta x} = \frac{y(x + \Delta x) - y(x)}{\Delta x} \quad \text{and} \quad \frac{dy}{dx} = \lim_{\Delta x \to 0} \frac{\Delta y}{\Delta x} = y'(x)$$

• *Distance* and *time* are sometimes represented using the definition of the derivative as:

$$f'(t) = \lim_{\Delta t \to 0} \frac{f(t + \Delta t) - f(t)}{\Delta t}$$

where Δt represents a small increment of time, such that the distance at some time (t + Δt) is represented by f(t + Δt). The distance at time t is represented by f(t) and the change in distance is $\Delta f = f(t + \Delta t) - f(t)$. The *average velocity* is the change in distance Δf divided by the change in time Δt, or Δf/Δt. The *instantaneous velocity* at a given time is found by shrinking Δt by taking the limit as Δt→0, which is f'(t) or df/dt. The *average slope* of the graph of f is Δf/Δt, and the *slope at some point* t on the graph of f is df/dt.

• For example, if a car is driving at a *constant velocity* of 65 mi/hr, the distance the car travels is given by f = d = vt. The distance traveled at any time t is f = vt, and at a later time (t + Δt) is v(t + Δt). The velocity can be represented by Δf/Δt. If $\Delta f = v \Delta t$ is substituted, velocity becomes $\Delta f/\Delta t = v \Delta t/\Delta t = v$, where $\lim_{\Delta t \to 0} \Delta f/\Delta t = df/dt = v$. In 1 hour, the car has traveled 65 miles = 65(1), at 2 hours, 130 miles, etc. Because the car is traveling at a constant velocity, Δf/Δt = df/dt = 65 mi/hr, and the limit is not required.

• Various notations are used to represent functions. It is important to understand what is being described and stay consistent with the notation within a given problem.

2.6 Slope of a Tangent Line and the Definition of the Derivative

• This section includes the slope of a tangent line and the definition of the derivative, and the equations for a tangent line, a secant line, and a normal line.

• In the graph of a function, the slope of a line drawn tangent to the curve through some point (a,f(a)) on the curve is the derivative of the function at point (a,f(a)). In other words, the *slope of the tangent* at point (a,f(a)) equals the derivative f'(a) at that point. (If a tangent line is vertical, its slope is undefined.) The slope of a tangent at a point measures the change in the curve at that point.

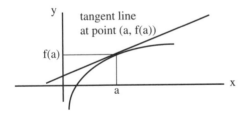

• The definition of the derivative can be used to prove that the slope of a line drawn tangent to a graph of a function at some point, is the derivative of the function at that point. Consider the two points on the curve (a,f(a)) and (a+h,f(a+h)). Tangent 1 is drawn through point (a,f(a)), tangent 2 is drawn through point (a+h,f(a+h)), and a "center line" is drawn through the two tangent points.

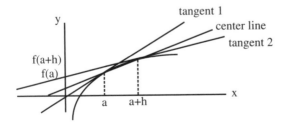

The slope of the center line through point (a,f(a)) and point (a+h,f(a+h)) represents the change in y over the change in x between the two points and is equal to:

$$\frac{f(a+h)-f(a)}{a+h-a} = \frac{f(a+h)-f(a)}{h}$$

If the value of the increment h between the two points is reduced, the value of h will approach zero, and the tangent 2 line through point (a+h,f(a+h)) will approach being equal to the tangent 1 line through point (a,f(a)). Therefore, the slope of the tangent at point (a,f(a)) equals the derivative f'(a):

$$f'(a) = \lim_{h \to 0} \frac{f(a+h)-f(a)}{h} = \text{the slope of tangent 1}$$

provided the limit exists.

Therefore, the slope of the tangent at (a,f(a)) is the derivative of f at point a.

• Note that the *derivative at a point* on a curve can be represented as either the slope of the tangent line to the curve at that point, or the slope of the curve at that point.

• The *equation for the tangent line* at y = f(x) and x = a is:

$$y - f(a) = f'(a)(x - a)$$

This can be derived as follows:

$$f'(a) \approx \frac{f(a+h)-f(a)}{h}$$

$$f'(a)(h) \approx f(a + h) - f(a)$$
$$f(a + h) \approx f'(a)(h) + f(a)$$

where if x − a = h and x = a + h,

$$f(a + h) \approx f'(a)(x - a) + f(a)$$
$$f(x) \approx f'(a)(x - a) + f(a)$$

using y = f(x),

$$y - f(a) = f'(a)(x - a)$$

This is used to *linearize* and estimate a region of f(x) close to x = a near the point (a,f(a)) on a curved function.

• The *secant line* is represented by a line drawn between two points on a curve. The equation for the secant line is:

$$y - f(x_1) = \frac{f(x_2) - f(x_1)}{x_2 - x_1}(x - x_1)$$

The *slope of the secant line*, $f'(x_1)$, is given by:

$$f'(x_1) = \lim_{x2 \to x1} \frac{f(x_2) - f(x_1)}{x_2 - x_1}$$

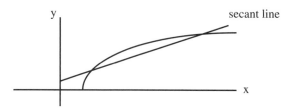

A *secant line* becomes a *tangent line* by letting $x = x_2$ approach $x = x_1$:

$$\lim_{x2 \to x1}\left[\frac{f(x_2) - f(x_1)}{x_2 - x_1}\right] = \lim_{x2 \to x1} \frac{\Delta f}{\Delta x} = \frac{df}{dx}$$

• Another important equation is the *equation for a line normal* or perpendicular to the tangent line on a curve at a given point. The slope of the tangent line and slope of a perpendicular line multiply to equal -1. If m is the tangent line and $-1/m$ is the normal line, then the equation for the normal line can be written:

$$y - y_1 = (-1/m)(x - x_1) \text{ or } y - f(a) = [-1/f'(a)](x - a)$$

2.7 Velocity, Distance, Slope, Area, and the Definition of the Derivative

• This section includes a summary of the relationship between velocity and distance; increasing velocity; constant velocity; and velocity, distance, and the area under a curve.

• The relationship between *distance traveled* f and *velocity* v is such that if f is known, v can be obtained, and if v is known, f can be obtained. Finding velocity from distance traveled involves *differentiation* and finding distance traveled from velocity involves *integration*. (Integration is discussed at length in Chapter 3.)

• Consider the graphs below of $f(t) = 3t^2$. Values of $v(t)$ can be calculated for various t values using the definition of the derivative by (a.) determining $f(t+h)$ and $f(t)$; (b.) subtracting $f(t)$ from $f(t+h)$; (c.) dividing $[f(t+h) - f(t)]$ by h; and (d.) taking the limit as h approaches zero, $\lim_{h \to 0}$.

Using these four steps for $f(t) = 3t^2$, $v(t)$ for $t = 2$ was calculated in Section 2.4 to be $v(2) = 12$. Using these same four steps for other values of t, $t = 0, 1, 3$, results in v values of $v(0) = 0$, $v(1) = 6$, and $v(3) = 18$. In summary:

$$f(0) = 0, \qquad v(0) = 0$$
$$f(1) = 3, \qquad v(1) = 6$$
$$f(2) = 12, \qquad v(2) = 12$$
$$f(3) = 27, \qquad v(3) = 18$$

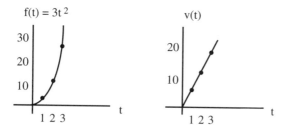

In this example, the *velocity is increasing with distance* and time. The slope of the curve drawn for $f(t)$ is equal to $v(t)$ at each point. Therefore, if a tangent line is drawn at each point for t, its slope is the velocity at that point or $v(t)$. Note that the slope of $v(t)$ is the acceleration.

• If *velocity v remains constant*, f will increase at that constant rate and $f = vt$. For example, if $v(t) = 6$ mi/h, then $f = 6t$. Therefore, for:

$$t = 1, \quad f = 6$$
$$t = 2, \quad f = 12$$
$$t = 3, \quad f = 18$$

$v(t)$ is constant at 6 since the slope of $f(t)$ is the same and equal to 6 at every point t.

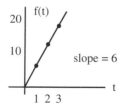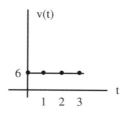

• The graph for v can be determined by calculating the slope of the f graph. When the slope of f is a *straight line*, the velocity is constant, and the graph of v(t) is a straight horizontal line at the constant slope value for f(t). When f is a *curve*, the slope must be calculated at each point by deter-mining the slope of a tangent line drawn at each point. The slope can be positive or negative depending on whether the velocity or rate-of-change is increasing or decreasing (accelerating or decelerating).

• To determine f from v graphically, it turns out that the *area under the curve (or line) of v gives f*. Therefore, the area under a graph of velocity represents the value for distance. This is discussed in Chapter 3. For a constant v the area is the rectangular region under v with height = v, width = t, and area = vt. This is consistent with f = vt. The area under v is a sum of areas at each t value that corresponds to distances at each time increment. For sloped or curved v graphs, the area can also be divided into incremental areas at each t value, where the velocity within each small increment is nearly constant.

• If a set of f values is listed, the differences between f's are v values. For example, if f = 1, 4, 8, 10, 12, then taking the difference between each f value results in a list of differences or v values, v = 3, 4, 2, 2. The sum of the differences in f values is $3 + 4 + 2 + 2 = 11$, which is equivalent to the difference between the first and last f value $12 - 1 = 11$. Therefore, the v values are the differences in f values at defined increments, and the *area under v* is the sum of the increments. The difference between the first and last value of f is:

$$\text{area} = f(t_{last}) - f(t_{first}), \text{ or,}$$
$$(f_1 - f_0) + (f_2 - f_1) + (f_3 - f_2) + (f_4 - f_3) + \ldots = f_n - f_0$$
$$= v_1 + v_2 + v_3 + \ldots + v_n$$

• For example, consider a sine wave type of pattern where the curve oscillates from positive to negative to positive, and so on. Values of f follow the pattern:

$$f = 0, 1, 1, 0, -1, -1, 0$$

where the differences which correspond to v values are:

$$v = 1, 0, -1, -1, 0, 1$$

The area is:

$$f(t_{last}) - f(t_{first}) = 0 = \text{sum of the differences (v values)}$$

Suppose more values are added:

$$f = 0, 1, 1, 0, -1, -1, 0, 1, 1, \text{the differences are:}$$
$$v = 1, 0, -1, -1, 0, 1, 1, 0$$
$$f(t_{last}) - f(t_{first}) = 1 = \text{sum of the differences (v values)}$$

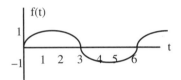 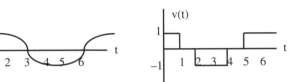

The area of v is the sum of the incremental positive and negative areas over a chosen interval of corresponding f and v values.

• For example, consider f = 1, 2, 3, 4, 5, 6. The differences are v = 1, 1, 1, 1, 1. The sum of the differences of the increments is: $f(t_{last}) - f(t_{first}) = 5$.

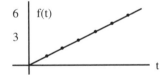 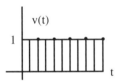

Slope of f = 1 = v, and area under v is $f(t_{last}) - f(t_{first}) = 5$.

2.8 Evaluating Derivatives of Constants and Linear Functions

• This section includes the derivative of a constant or a constant function and the derivative of a linear function. (The derivative of a constant multiplied by a variable is discussed in Section 2.10.)

• The derivative is the rate of change of something that is changing, there-fore the *derivative of a constant* is zero. The derivative of constant function f(x) = c is zero everywhere, because its graph is a horizontal line with a slope of zero everywhere. In the graph below, f(x) = c is a constant function with a slope of zero, therefore f'(x) = 0.

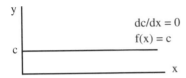

• The rate of change of a linear function is constant because for each change in x along the graph of a linear function, the corresponding changes in y are the same. The graph of a linear function is a straight line and the slope of a straight line is constant, therefore the *rate of change or derivative of a linear function* is constant.

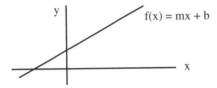

Remember the equation for a line is f(x) = mx + b, where the constant slope is m = derivative = f'(x). Calculating the derivative of f(x) = mx + b using the derivative formula (derived in the following section) is:

$$df(x)/dx = d(mx)/dx + db/dx = m + 0 = m$$

where m represents the constant slope of the straight line that represents the function.

2.9 Evaluating Derivatives Using the Derivative Formula

• This section introduces the derivative formula including its derivation.

• Evaluating derivatives using the definition of the derivative is labor-intensive. Instead, there is a shortcut formula used to evaluate derivatives. This *derivative formula* is:

$$\frac{d}{dx} x^n = nx^{n-1}$$

where x represents any variable and n is any number.

• If a constant a is multiplied by the variable x^n, the derivative formula becomes:

$$\frac{d}{dx} ax^n = anx^{n-1}$$

Note: The derivative formula is important and used frequently in calculus.

• The derivative formula can be derived from the definition of the derivative as follows:

Consider a function $f(x) = ax^n$ where a is some constant, x is a variable, and n is a number.

Substitute this function into the definition of the derivative:

$$\frac{d}{dx} ax^n = \lim_{h \to 0} \frac{a(x+h)^n - ax^n}{h}$$

Factor out the constant a:

$$\lim_{h \to 0} \frac{a((x+h)^n - x^n)}{h}$$

The $(x + h)^n$ term can be expanded using the *binomial expansion*:

$$(x + h)^n = x^n + nx^{n-1}h + \frac{n(n-1)x^{n-2}h^2}{2!} +$$

$$\frac{n(n-1)(n-2)x^{n-3}h^3}{3!} + \dots h^n$$

Because there are no h^2, h^3, etc., terms in the definition of the derivative, and the limit as h→0 will quickly remove these, write the binomial expansion excluding the h^2 and greater terms. This expansion becomes:

$$(x + h)^n = x^n + nx^{n-1}h$$

Substitute the expansion for $(x + h)^n$ into the definition of the derivative:

$$\lim_{h \to 0} \frac{a((x+h)^n - x^n)}{h} = \lim_{h \to 0} \frac{a((x^n + nx^{n-1}h) - x^n}{h}$$

Cancel the x^n terms:

$$\lim_{h \to 0} \frac{a(nx^{n-1}h)}{h}$$

Factor and cancel an h from each term:

$$\lim_{h \to 0} a(nx^{n-1})$$

Therefore, as $h \to 0$:

$$\frac{d}{dx} ax^n = anx^{n-1}$$

This is the derivative formula.

2.10 The Derivatives of a Variable, a Constant with a Variable, a Constant with a Function, and a Variable Raised to a Power

- In this section, the derivative formula is applied to

    $f(x) = x$, $f(x) = cx$, and $f(x) = cx^n$.

- The *derivative of an independent variable x with respect to itself* is one.

    $$\frac{d}{dx} x = 1 \times x^{1-1} = 1x^0 = 1$$

- The *derivative of a constant c times an independent variable x* with respect to x is equivalent to the constant times the derivative of the independent variable.

    $$\frac{d}{dx} cx = c \frac{d}{dx} x = c \times 1 \times x^{1-1} = c \times 1 \times x^0 = c$$

- The *derivative of a constant times a function* f(x) is equivalent to the constant times the derivative of the function.

    $$\frac{d}{dx} cf(x) = c \frac{d}{dx} f(x)$$

• By *multiplying a function by a constant*, the graph of that function will be affected by the constant. The graph will have the same general shape as it does without the multiple. However, it may have a larger or smaller amplitude, it may be elongated or narrower, it may be moved to the right or left or up or down, or if multiplied by a negative constant it may be flipped over the X-axis.

The slopes (or derivatives) of the curve of the graph of a function multiplied by a constant will be different at each point along the curve from the slopes of the curve of the original function. The change in the slopes will be proportional to the value of the constant.

For example, if a function is multiplied by the constant 2, the amplitude of the graph will be two times the amplitude of the original graph. Similarly, if a function is multiplied by the constant 1/2, the amplitude of the graph will be one-half the amplitude of the original graph.

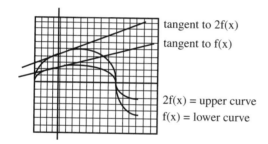

tangent to 2f(x)
tangent to f(x)

2f(x) = upper curve
f(x) = lower curve

• The *derivative of a function with the variable raised to a power* in the form cx^n evaluated using the derivative formula is:

$$\frac{d}{dx} cx^n = cnx^{n-1}$$

where c represents any constant number that may be multiplied with the function and n is an integer.

For example, $(d/dx)3x^2 = (3)(2)x^{2-1} = 6x$

If n is negative the formula becomes:

$$\frac{d}{dx} cx^{-n} = -cnx^{-n-1}$$

2.11 Examples of Differentiating Using the Derivative Formula

• This section includes using the derivative formula to evaluate simple functions and calculating the derivative of $1/x$ using the definition of the derivative as a comparison.

• The following are examples of using the *derivative formula* to evaluate derivatives:

$$\frac{d}{dx} x = 1 \times x^{1-1} = 1x^0 = 1$$

$$\frac{d}{dx} x^2 = 2 \times x^{2-1} = 2x^1 = 2x$$

$$\frac{d}{dx} x^{-3} = -3 \times x^{-3-1} = -3x^{-4}$$

$$\frac{d}{dx} x^{25} = 25 \times x^{25-1} = 25x^{24}$$

$$\frac{d}{dx} x^{0.25} = 0.25 \times x^{0.25-1} = 0.25x^{-0.75} = 0.25/x^{0.75} = 1/4x^{3/4}$$

$$\frac{d}{dx} x^{1/4} = 1/4 \times x^{(1/4)-1} = 1/4 \times x^{-3/4} = 1/4x^{3/4}$$

$$\frac{d}{dx} \sqrt{x} = \frac{d}{dx} x^{1/2} = 1/2 \times x^{(1/2)-1} = 1/2 \times x^{1/2-2/2} = (1/2)x^{-1/2}$$

$$= 1/(2x^{1/2}) = 1/(2\sqrt{x}) \text{ (Remember } x^n = 1/x^{-n} \text{ and } x^{-n} = 1/x^n.)$$

$$\frac{d}{dx} (1/x) = \frac{d}{dx} x^{-1} = -1 \times x^{-1-1} = -1x^{-2} = -1/x^2$$

$$\frac{d}{dx} (1/x^4) = \frac{d}{dx} x^{-4} = -4 \times x^{-4-1} = -4x^{-5} = -4/x^5$$

$$\frac{d}{dx}(2x) = 2x^{1-1} = 2x^0 = 2$$

$$\frac{d}{dx}(2x^2) = 2 \times 2x^{2-1} = 2 \times 2x^1 = 4x$$

$$\frac{d}{dx}2 = 0$$

• Using the derivative formula to evaluate $(d/dx)(1/x)$ resulted in $-1/x^2$. Evaluate this derivative using the *definition of the derivative* and compare the results:

Using $f'(x) = \lim_{\Delta x \to 0} \dfrac{f(x + \Delta x) - f(x)}{\Delta x}$, gives:

$$\lim_{\Delta x \to 0} \left(\left[\frac{1}{x + \Delta x} - \frac{1}{x} \right] \div \Delta x \right)$$

$$= \lim_{\Delta x \to 0} \left(\left[\frac{x}{x(x + \Delta x)} - \frac{(x + \Delta x)}{x(x + \Delta x)} \right] \div \Delta x \right)$$

$$= \lim_{\Delta x \to 0} \left(\left[\frac{x - (x + \Delta x)}{x(x + \Delta x)} \right] \div \Delta x \right)$$

$$= \lim_{\Delta x \to 0} \left(\left[\frac{-\Delta x}{x^2 + x\Delta x} \right] \div \Delta x \right) = \lim_{\Delta x \to 0} \left(\left[\frac{-1}{x^2 + x\Delta x} \right] \right)$$

as $\Delta x \to 0$, this approaches $-1/x^2$, which is the same answer obtained using the derivative formula.

2.12 Derivatives of Powers of Functions

• This section includes evaluating powers of functions using the derivative formula, comparing results with the definition of the derivative and graphs.

• To evaluate the *derivatives of powers of functions*, the derivative formula can be applied to the whole function, which is then multiplied with the derivative of what is inside. This is known as the *chain rule* and is discussed in Section 2.19.

$$\frac{d}{dx}(f(x))^2 = 2 \times f(x) \times \frac{d}{dx}f(x)$$

$$\frac{d}{dx}(f(x))^3 = 3 \times (f(x))^2 \times \frac{d}{dx}f(x)$$

$$\frac{d}{dx}(f(x))^4 = 4 \times (f(x))^3 \times \frac{d}{dx}f(x)$$

$$\frac{d}{dx}(f(x))^n = n \times (f(x))^{n-1} \times \frac{d}{dx}f(x)$$

• Results obtained using this method can be compared with results obtained using the *definition of the derivative*. For example, if $(d/dx)(f(x))^2$:

$$\frac{\Delta f}{\Delta x} = \frac{(f(x+\Delta x))^2 - (f(x))^2}{\Delta x}$$

Remember the factored form of $(x^2 - y^2)$ is $(x + y)(x - y)$. Using this for Δf:

$$\frac{\Delta f}{\Delta x} = \frac{(f(x+\Delta x)+f(x))(f(x+\Delta x)-f(x))}{\Delta x}$$

$$= (f(x+\Delta x)+f(x))\frac{f(x+\Delta x)-f(x)}{\Delta x}$$

therefore, $\dfrac{df}{dx} = \lim_{\Delta x \to 0}\dfrac{\Delta f}{\Delta x} = 2f(x)\dfrac{df}{dx}$

• Following are *graphs of functions raised to a power and their derivatives.*

Example: If $f(x) = x^3$, using the derivative formula results in $3x^2$. The graph of x^3 and its derivative is represented as:

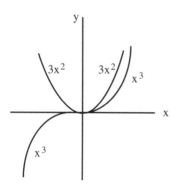

Example: If f(x) = x², using the derivative formula results in 2x. The graph of x² and its derivative is:

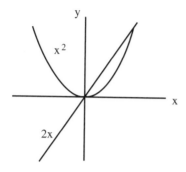

Because the derivative of x² is 2x, which is a linear function, the graph of the derivative is a straight line with a constant slope of 2.

2.13 Derivatives of aˣ, eˣ, and ln x

• This section includes calculating derivatives along a curve of aˣ, demonstrating that the derivative of eˣ is eˣ, the relationship between the derivative of aˣ and the natural logarithm, and the derivative of the natural logarithm and of functions that involve the natural logarithm.

• Graphs of *exponential functions* in the form f(x) = eˣ and f(x) = aˣ, depict that for negative values of x, f(x) increases slowly and for positive values of x, f(x) increases faster.

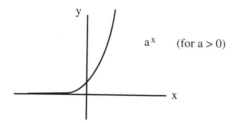

For real values of x, the graph of the derivative of f(x) = aˣ exists above the X-axis.

• Consider $f(x) = a^x$ when $a = 2$. The curve of $f(x) = 2^x$ can be plotted by selecting x values and solving for $f(x)$:

$$x = -3, f(x) = 2^{-3} = 1/2^3 = 1/8$$
$$x = -2, f(x) = 2^{-2} = 1/2^2 = 1/4$$
$$x = -1, f(x) = 2^{-1} = 1/2^1 = 1/2$$
$$x = 0, f(x) = 2^0 = 1$$
$$x = 1, f(x) = 2^1 = 2$$
$$x = 2, f(x) = 2^2 = 4$$
$$x = 3, f(x) = 2^3 = 8$$
$$x = 4, f(x) = 2^4 = 16$$

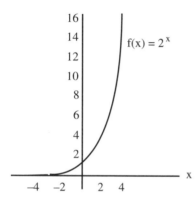

It is possible to *calculate the derivative* for $f(x) = 2^x$ at points along the curve using the definition of the derivative. This can be developed by first applying the *definition of the derivative* to the general case of a^x and developing a general formula. Begin with:

$$f'(x) = \lim_{h \to 0} \frac{f(a+h) - f(a)}{h}$$

For $f(x) = a^x$, $f'(x) = \lim_{h \to 0} \frac{a^{x+h} - a^x}{h} = \lim_{h \to 0} \frac{a^x a^h - a^x}{h}$

$$= \lim_{h \to 0} \frac{a^x(a^h - 1)}{h} = a^x \lim_{h \to 0} \frac{a^h - 1}{h}$$

therefore, $f'(x) = a^x \lim_{h \to 0} \dfrac{a^h - 1}{h}$

This formula shows that the *derivative of* a^x is equal to a^x multiplied by the limit of a constant. This expression for $f'(x)$ can then be used to calculate values of the slope and, thus, the derivative of a^x at various points along the curve. Using this formula to calculate the derivative for $f(x) = 2^x$, where a = 2, first select x values along the curve such as $x = -1$, $x = 0$, and $x = 1$, and apply the formula:

$$f'(-1) = 2^{-1} \lim_{h \to 0} \frac{2^h - 1}{h} = 1/2 \lim_{h \to 0} \frac{2^h - 1}{h}$$

$$f'(0) = 2^0 \lim_{h \to 0} \frac{2^h - 1}{h} = 1 \lim_{h \to 0} \frac{2^h - 1}{h}$$

$$f'(1) = 2^1 \lim_{h \to 0} \frac{2^h - 1}{h} = 2 \lim_{h \to 0} \frac{2^h - 1}{h}$$

In general, for $f'(x) = 2^x \lim_{h \to 0} \frac{2^h - 1}{h}$

Taking the limit as h→0, by choosing small h values to see what the value of $(2^h - 1)/h$ approaches:

h = −0.001, the limit = 0.692907
h = +0.001, the limit = 0.693387
h = −0.0002, the limit = 0.693099
h = +0.0002, the limit = 0.693195
h = −0.0001, the limit = 0.693123
h = +0.0001, the limit = 0.693171

Therefore, as h→0, $(2^h - 1)/h$ approaches 0.6931.

The derivative of $f(x) = 2^x$ can be calculated using:

$(d/dx)2^x = 2^x(0.6931)$

It is then possible to use this resulting formula to calculate the derivative at various points along the curve of 2^x for different x values such as $x = -1, 0, 1$:

$f'(-1) = 2^{-1}\lim_{h \to 0}(2^h - 1)/h = 2^{-1}(0.6931) = 1/2(0.6931) = 0.346$
$f'(0) = 2^0\lim_{h \to 0}(2^h - 1)/h = 2^0(0.6931) = 1(0.6931) = 0.693$
$f'(1) = 2^1\lim_{h \to 0}(2^h - 1)/h = 2^1(0.6931) = 2(0.6931) = 1.386$

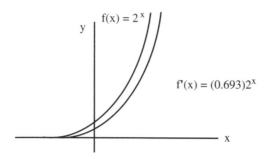

- Consider $f(x) = a^x$ where $a = e$. The general expression
$f'(x) = a^x \lim_{h \to 0}(a^h - 1) / h$ developed in the preceding paragraphs
can be used to show that the derivative of e^x is e^x. To show that the
derivative of e^x is equal to the original function e^x, first notice that the
quantity $\lim_{h \to 0}(a^h - 1) / h$ does not depend on the value of x and is therefore a constant for each unique value of a. It was calculated in the preceding
paragraphs that this quantity of the limit is 0.6931 when $a = 2$ and therefore
$f'(x) = 2^x(0.6931)$.

Similarly, it can be calculated using the same process that the quantity of
the limit is 1.0986 when $a = 3$, and therefore $f'(x) = 3^x(1.0986)$. Notice
that when $a = 2$, the quantity of the limit is less than 1 and the derivative
of a^x is slightly less than the original function a^x. Also, when $a = 3$, the
quantity of the limit is greater than 1 and the derivative of a^x is slightly
greater than the original function a^x.

Expanding on this and using these observations in combination with the
fact that $e = 2.7818$, the quantity of the limit must be equal to 1. This is
consistent with the fact that the derivative of e^x must be equal to the
original function e^x.

It is possible to find the value of a when $a = e$, which is the number that
e represents, using this quantity of the limit where it is equal to 1:

$\lim_{h \to 0}(a^h - 1) / h = 1$

First isolate a by considering the quantity of the limit at small h values:

$(a^h - 1) / h \approx 1$
$a^h - 1 \approx h$
$a^h \approx h + 1$
$a \approx (h + 1)^{1/h}$

Select small values for h as h→0:

    h = 0.0001, a ≈ 2.7181
    h = −0.0001, a ≈ 2.7184
    h = 0.00001, a ≈ 2.7183
    h = −0.00001, a ≈ 2.7183

These values of a = e converge to a = e = 2.71828...

By substituting e = 2.718 and choosing small h values, it can be shown that: $\lim_{h \to 0}(e^h - 1)/h = 1$

Therefore, $(d/dx)e^x = e^x$

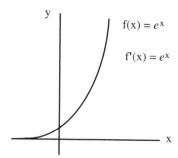

Note that the slope at x = 0 is one.

- Note that e can be expressed in the following two forms:

$$e = \lim_{n \to \infty}\left(1 + \frac{1}{n}\right)^n$$

$$e = \frac{1}{0!} + \frac{1}{1!} + \frac{1}{2!} + \frac{1}{3!} + \frac{1}{4!} + \ldots$$

- The exponential function is related to the *natural logarithm* such that $\ln(e^x)$ = x and $e^{\ln x}$ = x. In the examples of the derivatives of a^x where a = 2 and a = 3, the quantities of the limits are 0.6931 and 1.0986, respectively. These values are natural logarithms, such that 0.6931 ≈ ln 2 and 1.0986 ≈ ln 3. Therefore, using the formula developed in the preceding paragraphs:

$$\frac{d}{dx} 2^x = 2^x(0.6931) = 2^x(\ln 2)$$

$$\frac{d}{dx} 3^x = 3^x(1.0986) = 3^x(\ln 3)$$

Similarly, the following derivatives can be written:

$$\frac{d}{dx} a^x = a^x(\ln a)$$

$$\frac{d}{dx} e^x = e^x(\ln e)$$

• The *derivative of the natural logarithm* of x, or ln x, is 1/x. This can be verified using exponential form and the fact that $x = e^{\ln x}$ when $x > 0$ as follows:

Begin by differentiating both sides of $x = e^{\ln x}$:

$$\frac{d}{dx} x = \frac{d}{dx} e^{\ln x}$$

Using the chain rule (described in Section 2.19.) where $f(g(x))' = f'(g(x))(g'(x))$, differentiate the right side $e^{\ln x}$, or $e^{f(x)}$:

$$\frac{d}{dx} e^{f(x)} = e^{f(x)}(f'(x))$$

Substitute $f(x) = \ln x$:

$$\frac{d}{dx} e^{\ln x} = e^{\ln x} \frac{d}{dx} (\ln x) = x \frac{d}{dx} (\ln x)$$

Set equal to the left side:

$$\frac{d}{dx} x = x \frac{d}{dx} (\ln x)$$

$$1 = x \frac{d}{dx} (\ln x)$$

$$1/x = \frac{d}{dx} (\ln x), \text{ for } x > 0 \text{ or } |x| \neq 0$$

Therefore, $1/x$ = the slope of the graph of ln x.

• Following are examples of *derivatives* that involve the *natural logarithm*:

$$\frac{d}{dx}(\ln(f(x))) = (\frac{1}{f(x)})(f'(x)), \text{ for } f(x) > 0 \text{ and } x \neq 0$$

$$\frac{d}{dx}(\ln(x^2)) = 2x(1/x^2) = 2/x$$

$$\frac{d}{dx}(\ln 2x) = 2(1/2x) = 1/x, \text{ where 2 does not affect the slope.}$$

$$\frac{d}{dx}(\ln(x^3 + 3)) = 3x^2/(x^3 + 3)$$

$$\frac{d}{dx}(\ln(\sin x)) = \cos x(\frac{1}{\sin x}) = \frac{\cos x}{\sin x}$$

2.14 Applications of Exponential Equations

• This section includes the general equation for growth and decay, an example of a bacteria population, and an equation for compound interest.

• There are many questions in science, finance, etc., that can be answered using *growth and decay* models. For example, the rate of change of growth of a population is often proportional to the size of the population at a given point in time. The general *equation for exponential growth and decay* is given by the derivative:

$$dy(t)/dt = ky(t), \text{ where } y(t) = ce^{kt} \text{ and } \frac{dy(t)}{dt} = kce^{kt}$$

y(t) represents what is growing or decaying at time t (e.g., the size of a population or the amount of a radioactive substance);
t represents time;
c represents the initial amount (e.g., mass); and
k is a constant of proportionality representing rate of growth or decay, such that when k > 0 the population is growing at an exponential rate, and when k < 0 the population is decreasing at an exponential rate.

Differentiating y(t) gives:

$$\frac{dy(t)}{dt} = \frac{d}{dt}(ce^{kt}) = (ce^{kt})(\frac{d}{dt}kt) = kce^{kt} = ky(t)$$

Therefore, for any function where $dy(t)/dt = ky(t)$, then $y(t) = ce^{kt}$, for some number c.

Note that at $t = 0$, $y(0) = ce^{k(0)} = c$

• **Example:** Suppose a bacterial population (culture) begins with 100,000 bacteria and in 20 days has 200,000 bacteria. If the population will double in 20 days, how big will it be in 15 days and what is its rate of change in 20 days?

y(t) represents the size of the population in t days

$y(t) = ce^{kt}$

First determine c and k:

At $t = 0$, $y(0) = ce^{k(0)} = c = 100,000$

To find k, at $t = 20$, $y(20) = 100,000\ e^{k(20)} = 200,000$

Rearranging gives: $e^{k(20)} = 2$

Take logarithm of both sides:

ln $e^{k(20)}$ = ln 2
$k(20)$ = ln 2
$k = (ln\ 2)/20$

Substitute c and k into the equation for y(t):

$y(t) = (100,000)e^{((ln2)/20)\ t}$

At $t = 15$:

$y(15) = 100,000\ e^{((ln\ 2)/20)(15)} = (100,000)2^{15/20} =$
$y(15) = 168,179$ bacteria in 15 days

The rate of change of the population at 20 days is:

$$\frac{dy(20)}{dt} = ky(20) = (\ln 2/20)(100{,}000)e^{(t \ln 2)/20}$$

$$= (5{,}000)(\ln 2)\, e^{\ln 2} = (5{,}000)(\ln 2)(2) = (10{,}000)\ln 2$$
$$= 6{,}931 \text{ bacteria/day}$$

• For examples involving interest on money, the amount of interest earned at a continuously compounded fixed rate will be dependent on the amount of money, and will increase with increasing amounts of money. The growth-decay equation applies to this as well:

$$y(t) = ce^{kt}$$

where $y(t)$ represents the amount of money at time t;
c represents the starting amount;
k represents the continuous or instantaneous growth rate (interest rate). Note that k is not the same as the annual growth rate.

• Continuously applied compounded interest is sometimes represented by:

$$P(1 + (i/n))^t$$

where P represents principal;
t represents number of time periods where amount is checked;
i represents yearly interest rate (decimal form);
n represents number of time periods per year that equal divisions of the proportional amount of interest is paid; and
i/n represents annual growth rate.

At one year the amount is $P(1 + (i/n))^n$.

As number of times per year interest is paid increases (such that it is "continuously compounded"):

$$\lim_{n \to \infty} P(1 + (i/n))^n = P \lim_{n \to \infty} (1 + (i/n))^n$$

Substitute "in" for n:

$$P \lim_{n \to \infty} (1 + (i/in))^{in} = P \lim_{n \to \infty} (1 + (1/n))^{in}$$

Because e can be expressed as $e = \lim_{n \to \infty}(1 + (1/n))^n$, then,

   $P \lim_{n \to \infty}(1 + (1/n))^{in}$ becomes:

Pe^i = continuously compounded interest.

2.15 Differentiating Sums, Differences, and Polynomials

• This section includes differentiating sums and differences of functions and differentiating polynomial functions.

• To *differentiate functions that are added or subtracted*, differentiate each function separately, then add or subtract the resulting functions. If one of the functions contains a polynomial, differentiate the polynomial term by term.

• The sum and difference of the two functions $f(x)$ and $g(x)$ can be differentiated as follows:

$$\frac{d}{dx}[f(x) + g(x)] = \frac{d}{dx}f(x) + \frac{d}{dx}g(x)$$

$$\frac{d}{dx}[f(x) - g(x)] = \frac{d}{dx}f(x) - \frac{d}{dx}g(x)$$

• **Example:** If $f(x) = 2x^2$ and $g(x) = x^3 + 3$, find:

$$\frac{d}{dx}[f(x) + g(x)]$$

$$\frac{d}{dx}[f(x) + g(x)] = \frac{d}{dx}[(2x^2) + (x^3 + 3)]$$

$$= \frac{d}{dx}(2x^2) + \frac{d}{dx}(x^3 + 3) = \frac{d}{dx}2x^2 + \frac{d}{dx}x^3 + \frac{d}{dx}3$$

$$= 4x + 3x^2 + 0 = 4x + 3x^2$$

- **Example:** Find:

$$\frac{d}{dx} [\sin x + \cos x]$$

$$\frac{d}{dx} \sin x + \frac{d}{dx} \cos x = \cos x - \sin x$$

- *Polynomial functions* can be differentiated by differentiating each term separately. For example, if x_1, x_2, $-x_3$, and x_4 each represent a term, then:

$$\frac{d}{dx} (x_1 + x_2 - x_3 + x_4) = \frac{d}{dx} x_1 + \frac{d}{dx} x_2 - \frac{d}{dx} x_3 + \frac{d}{dx} x_4$$

Differentiating a polynomial is similar to differentiating a sum or difference:

$$\frac{d}{dx} (f + g + c) = \frac{d}{dx} f + \frac{d}{dx} g + \frac{d}{dx} c$$

- For example, differentiate the following polynomial function term by term using the derivative formula:

$$\frac{d}{dx} (x^3 + 4x^2 + 7x + 9) = \frac{d}{dx} x^3 + \frac{d}{dx} 4x^2 + \frac{d}{dx} 7x + \frac{d}{dx} 9$$

$$= 3x^2 + (4)(2x) + (7)(1) + 0 = 3x^2 + 8x + 7$$

2.16 Taking Second Derivatives

- This section includes the definition of the second derivative, an example, and notation.

- In general, the *second derivative* of a function involves taking the derivative of the function that results after the first derivative is taken. The second derivative is the rate of change of the rate of change. For example, *velocity* v is the rate at which the position of something is changing with respect to time, and *acceleration* a is the rate at which the velocity is changing with respect to time.

    v = dx/dt and a = dv/dt

therefore, a = dv/dt = d²x/dt².

A positive value of a reflects acceleration, and a negative value of a reflects deceleration.

- If $y = f(x)$, the first derivative is $f'(x) = dy/dx$, then the second derivative is $f''(x) = \dfrac{d}{dx}\left(\dfrac{dy}{dx}\right) = \dfrac{d^2y}{dx^2}$

- The second derivative provides information about change in slope, the rate of change of something that is changing, such as the growth rate of a population, and is used in minima and maxima problems. (See Section 2.27.)

- To find the second derivative of a function, differentiate the original function first, then differentiate the result. For example, for the polynomial function in Section 2.15,

   $f(x) = (x^3 + 4x^2 + 7x + 9)$, the first derivative is $3x^2 + 8x + 7$.

To find the second derivative, differentiate this resulting function:

$$\frac{d}{dx}(3x^2 + 8x + 7) = \frac{d}{dx} 3x^2 + \frac{d}{dx} 8x + \frac{d}{dx} 7$$

$$= 6x + 8 + 0 = 6x + 8$$

- *Notation for multiple derivatives* is:

For the second derivative:

$$\frac{d^2f(x)}{dx^2}, \frac{d^2}{dx^2} f(x),\ f^2(x),\ f''(x),\ \mathbf{D}^2f(x),\ \mathbf{D}_x^2f(x)$$

For the nth derivative:

$$\frac{d^nf(x)}{dx^n}, \frac{d^n}{dx^n} f(x),\ f^n(x),\ \mathbf{D}^nf(x),\ \mathbf{D}_x^nf(x)$$

2.17 Derivatives of Products: The Product Rule

- This section includes the definition of the product rule, derivation of the product rule, and the product rule for multiple products.

• The *product rule* can be used to differentiate the product of two functions. The product rule applied to the product of the two functions f(x) and g(x) is:

$$\frac{d}{dx}(f(x))(g(x)) = (\frac{d}{dx}f(x))(g(x)) + (f(x))(\frac{d}{dx}g(x))$$

Using shorthand notation the product rule is written:

$$(fg)' = f'g + fg'$$

Note: The formula for the product rule is important and used frequently in calculus.

• The product rule can be developed using the definition of the derivative given by:

$$\frac{df(x)}{dx} = \lim_{h \to 0} \frac{f(x+h) - f(x)}{h}$$

which can be written for two functions as:

$$f'(x)g'(x) = \lim_{h \to 0} \frac{f(x+h)g(x+h) - f(x)g(x)}{h}$$

Using the Δ notation and the definition of the derivative, a small change in f, or Δf, is $f(x + \Delta x) - f(x)$ and a small change in g, or Δg is $g(x + \Delta x) - g(x)$, such that:

$$f'(x)g'(x) = \lim_{\Delta x \to 0} \frac{f(x+\Delta x)g(x+\Delta x) - f(x)g(x)}{\Delta x}$$

Because, $\Delta f = f(x + \Delta x) - f(x)$

rearrange:

$$f(x + \Delta x) = \Delta f + f(x)$$

Similarly for g:

$$\Delta g = g(x + \Delta x) - g(x)$$

rearrange:

$$g(x + \Delta x) = \Delta g + g(x)$$

Then substitute:

$$f'(x)g'(x) = \lim_{\Delta x \to 0} \frac{(\Delta f + f(x))(\Delta g + g(x)) - f(x)g(x)}{\Delta x}$$

$$= \lim_{\Delta x \to 0} \frac{\Delta f \cdot \Delta g + \Delta f \cdot g(x) + \Delta g \cdot f(x) + f(x)g(x) - f(x)g(x)}{\Delta x}$$

$$= \lim_{\Delta x \to 0} \frac{\Delta f \cdot \Delta g + \Delta f \cdot g(x) + \Delta g \cdot f(x)}{\Delta x}$$

$$= \lim_{\Delta x \to 0} \left(\frac{\Delta f \cdot \Delta g}{\Delta x} + \frac{\Delta f \cdot g(x)}{\Delta x} + \frac{\Delta g \cdot f(x)}{\Delta x} \right)$$

$$= \lim_{\Delta x \to 0} \frac{\Delta f \cdot \Delta g}{\Delta x} + \lim_{\Delta x \to 0} \frac{\Delta f \cdot g(x)}{\Delta x} + \lim_{\Delta x \to 0} \frac{\Delta g \cdot f(x)}{\Delta x}$$

Multiply the first term by $\Delta x / \Delta x$ and remember that:

$$\lim_{\Delta x \to 0} \frac{\Delta y}{\Delta x} = \frac{dy}{dx}:$$

$$\lim_{\Delta x \to 0} \frac{\Delta f \cdot \Delta g}{\Delta x} \frac{\Delta x}{\Delta x} = \lim_{\Delta x \to 0} \frac{\Delta f \cdot \Delta g}{\Delta x \cdot \Delta x} \Delta x = \frac{df}{dx} \frac{dg}{dx} \cdot 0 = 0$$

results in:

$$f'(x)g'(x) = 0 + \lim_{\Delta x \to 0} \frac{\Delta f}{\Delta x} g(x) + \lim_{\Delta x \to 0} \frac{\Delta g}{\Delta x} f(x)$$

$$= \frac{df}{dx} g(x) + \frac{dg}{dx} f(x) = f'(x)g(x) + g'(x)f(x)$$

which is the product rule.

- **Example:** If $f(x) = x^2$ and $g(x) = x^3$, find:

  $$\frac{d}{dx} (f(x))(g(x)) \text{ using the product rule.}$$

First evaluate f' and g':

$$f'(x) = 2x, g'(x) = 3x^2$$

Apply the product rule:

$$\frac{d}{dx}(x^2)(x^3) = (\frac{d}{dx}(x^2))(x^3) + (x^2)(\frac{d}{dx}(x^3))$$

$$= (2x)(x^3) + (x^2)(3x^2) = 2x^4 + 3x^4 = 5x^4$$

• The *product rule* can be applied to find the derivative of a *function raised to the second power*. For example, $(x^3 + x^2)^2$ can be treated as the product of $(x^3 + x^2)(x^3 + x^2)$.

• An extension of the product rule can be applied to *derivatives of multiple products*.

For two functions f and g: $(fg)' = f'g + fg'$
For three functions f, g, and h: $(fgh)' = f'gh + fg'h + fgh'$
For four functions f, g, h, and p:
$(fghp)' = f'ghp + fg'hp + fgh'p + fghp'$

2.18 Derivatives of Quotients: The Quotient Rule

• This section includes the definition of the quotient rule, a proof of the quotient rule, and an example.

• The *quotient rule* can be applied to evaluate derivatives of quotients of functions. For the functions $f(x)$ and $g(x)$ the quotient rule is:

$$\frac{d}{dx}\frac{f(x)}{g(x)} = \frac{f'(x)g(x) - f(x)g'(x)}{g(x)^2}$$

Or equivalently:

$$\left(\frac{f}{g}\right)' = \frac{f'g - fg'}{g^2}$$

Note: The formula for the quotient rule is important and used frequently in calculus.

• To prove the quotient rule, first let quotient, $Q = f/g$, then by rearranging, $f = Qg$.

Apply the product rule to $f = Qg$:

$$f' = Q'g + Qg'$$

Substitute $Q = f/g$:

$$f' = Q'g + (f/g)g'$$

Solve for Q':

$$Q'g = f' - (f/g)g'$$

$$Q' = \frac{f' - (f/g)g'}{g}$$

Multiply both sides by (g/g):

$$(g/g)Q' = \frac{(g/g)f' - (g/g)(f/g)g'}{g}$$

$$Q' = (f/g)' = \frac{f'g - fg'}{g^2} \text{ which is the quotient rule.}$$

• **Example:** Use the quotient rule to find the derivative of the quotient of $f(x) = x^2$ and $g(x) = x^3$, or: $(d/dx)(x^2/x^3)$.

To find the derivative of the quotient, evaluate $f'(x)$ and $g'(x)$:

$$f'(x) = 2x, \ g'(x) = 3x^2$$

Substitute $f(x)$, $g(x)$, $f'(x)$, $g'(x)$ into the quotient rule:

$$\frac{d}{dx}\frac{x^2}{x^3} = \frac{2x(x^3) - (x^2)3x^2}{(x^3)^2} = \frac{2x^4 - 3x^4}{x^6} = \frac{-x^4}{x^6} = -1/x^2$$

Therefore, $(d/dx)(x^2/x^3) = -1/x^2$.

Note that this simple example is used so that the result can be verified by first simplifying the expression x^2/x^3 to $1/x$, then differentiating:

$$\frac{d}{dx} 1/x = \frac{d}{dx} x^{-1} = -1x^{-1-1} = -x^{-2} = -1/x^2$$

2.19 The Chain Rule for Differentiating Complicated Functions

• This section includes the definition of the chain rule, examples using the chain rule, the chain rule applied to reciprocal functions, and functions raised to a power.

- The *chain rule* can be used to *differentiate composite functions* in which variables depend on other variables. Consider the function f that depends on the variable u, but u depends on the variable x. In other words, f is a function of u, and u is a function of x. The chain rule is written:

$$\frac{d}{dx} f(u(x)) = \frac{d}{du} f(u) \times \frac{d}{dx} u(x) = (f'(u))(u'(x))$$

where the derivative exists at u(x) and at f(u(x)).

Note: The formula for the chain rule is important and used frequently in calculus.

- The chain rule is the derivative of the outer function times the derivative of the function inside. It is important to identify the outer and inner functions in a *composite* function that is to be differentiated. In the function f(u(x)), f is the outside function and u is the inside function. Similarly, in the function f(g(x)), f is the outside function and g is the inside function.

- The chain rule can be derived as follows:

  If z = u(x) and y = f(z), then y = f(u(x)).

In this function, u(x) is determined by x and f(z) is determined by z. A small change in x, Δx, will cause a small change in z, Δz, which will cause a small change in y, Δy.

Therefore, $\dfrac{\Delta y}{\Delta x} = \dfrac{\Delta y}{\Delta z} \dfrac{\Delta z}{\Delta x}$

Because,

$$\frac{dy}{dx} = \lim_{\Delta x \to 0} \frac{\Delta y}{\Delta x}, \frac{dy}{dz} = \lim_{\Delta z \to 0} \frac{\Delta y}{\Delta z} \text{ and } \frac{dz}{dx} = \lim_{\Delta x \to 0} \frac{\Delta z}{\Delta x},$$

taking the limits of each term results in:

$$\frac{dy}{dx} = \frac{dy}{dz} \frac{dz}{dx} = \frac{d}{dx} f(u(x)) = (f'(u))(u'(x))$$

which is the chain rule,

where $\dfrac{dy}{dz} \to f'(u)$ and $\dfrac{dz}{dx} \to u'(x)$.

- **Example:** If $f(u) = (u(x))^2$ and $u(x) = x^3$, apply the chain rule:

$$\frac{d}{dx}f[u(x)] = \frac{d}{du}(u(x))^2 \times \frac{d}{dx}x^3 = 2u(x) \times 3x^2$$

Substitute $u(x) = x^3$:

$$2x^3 \times 3x^2 = 6x^5$$

Therefore, $\dfrac{d}{dx} f(u(x)) = 6x^5$.

This is a simple example and is used so that the result can be verified by substituting $u(x) = x^3$ directly into $f(u) = (u(x))^2$ first and then evaluating the derivative.

$$f(u) = (u(x))^2 = (x^3)^2 = x^6$$

$$\frac{d}{dx}f(u(x)) = \frac{d}{dx}x^6 = 6x^5$$

- **Example:** The chain rule can be used to break complex functions into two simpler functions. Consider the derivative $(d/dx)f(x) = (d/dx)[(x^2 + x)^3]$.

This can be simplified using the chain rule.

First let $f(u) = (u(x))^3$ and $u(x) = (x^2 + x)$.

Using the chain rule, $\dfrac{d}{dx}f(u(x)) = \dfrac{d}{du}f(u) \times \dfrac{d}{dx}u(x)$,

substitute for $f(u)$ and $u(x)$:

$$\frac{d}{dx}f[u(x)] = \frac{d}{du}(u(x))^3 \times \frac{d}{dx}(x^2 + x)$$

Differentiate:

$$\frac{d}{dx}f[u(x)] = 3 \times (u(x))^2 \times (2x + 1)$$

Substitute $u(x) = (x^2 + x)$:

$$\frac{d}{dx}f[u(x)] = 3 \times (x^2 + x)^2 \times (2x + 1)$$

$$= 3 \times (x^2 + x) \times (x^2 + x) \times (2x + 1)$$

Multiply the first two binomials:

$$= 3 \times (x^4 + x^3 + x^3 + x^2) \times (2x + 1)$$

Combine like terms:

$$= 3 \times (x^4 + 2x^3 + x^2) \times (2x + 1)$$

Multiply the binomial and the trinomial:

$$= 3 \times [(2x^5 + 4x^4 + 2x^3) + (x^4 + 2x^3 + x^2)]$$

Combine like terms, then multiply the 3:

$$= 3 \times [2x^5 + 5x^4 + 4x^3 + x^2] = 6x^5 + 15x^4 + 12x^3 + 3x^2$$

Therefore, $\dfrac{d}{dx}[(x^2 + x)^3] = 6x^5 + 15x^4 + 12x^3 + 3x^2$.

- **Example:** If $y = (\cos x)^3$, what is dy/dx?

Using the chain rule:

$$\frac{d}{dx}y = \frac{d}{dx}(\cos x)^3 \times \frac{d}{dx}\cos x = (3 \cos x)^2 \times (-\sin x)$$

$$= -3 \cos^2 x \sin x$$

- **Example:** If $y = \exp\{x^{1/2}\}$ what is dy/dx?

("exp" represents e)

$$\frac{d}{dx}y = \frac{d}{dx}(\exp\{x^{1/2}\}) \times \frac{d}{dx}x^{1/2} = (\exp\{x^{1/2}\})((1/2)x^{(1/2-2/2)})$$

$$= (\exp\{x^{1/2}\})((1/2)x^{-1/2}) = \frac{\exp\{x^{1/2}\}}{2x^{1/2}}$$

- The *derivative of a reciprocal function* 1/f(x) can be solved using the chain rule as follows:

$$\frac{d}{dx}(1/f(x)) = \frac{d}{dx}(f(x))^{-1} \times \frac{d}{dx}f(x)$$

$$= (-1)(f(x))^{-2} \times \frac{d}{dx}f(x) = \frac{-df(x)/dx}{(f(x))^2}$$

- The *derivative of a function raised to a power* can be solved using the chain rule as follows:

$$\frac{d}{dx}(f(x))^n = \frac{d}{dx}(f(x))^n \times \frac{d}{dx}f(x)$$

$$= n \times (f(x))^{n-1} \times \frac{d}{dx}f(x) = n(f(x))^{n-1} \times f'(x)$$

2.20 Rate Problem Examples

- This section includes two examples of rate problems.

- *Rate problems* are common in calculus and are used to determine rates of movement or change in some parameter with respect to time.

- For example, consider circular waves caused from an object tossed into a pool of water where the circular ripples increase in diameter at a rate of 10 cm/second, at a diameter of 10 cm. How fast are the circumference and area of the circular ripples increasing at radius = 5 cm?

Given c = circumference = $2\pi r$, a = area = πr^2, and the rate of change is 10 cm/second, which is (dr(t)/dt), then at radius r = 5 cm the change in circumference is:

$$\frac{d}{dt}c(t) = \frac{d}{dt}2\pi r(t) = 2\pi \frac{d}{dt}r(t) = 2\pi(10) = 20\pi \text{ cm/s}$$

At r = 5 cm the change in area is:

$$\frac{d}{dt}a(t) = \frac{d}{dt}\pi(r(t))^2 = \pi \frac{d}{dt}(r(t))^2 = 2\pi r(t)\frac{d}{dt}r(t)$$

$$= 2\pi(5)\frac{d}{dt}r(t) = 10\pi(10) = 100\pi \text{ cm/s}$$

- Another example of the "rate" problem involves the stretching of a right triangle:

If side A is fixed at 5 cm and side B is stretching at a constant rate of 2 cm/s so that (dB/dt) = 2 cm/s, then what is the rate of change of C, (dC/dt), when B is at 5 cm?

Using the Pythagorean Theorem $A^2 + B^2 = C^2$:

$$\frac{d}{dt} C^2 = \frac{d}{dt} [A^2 + B^2] = \frac{d}{dt} [5^2 + B^2]$$

Taking derivatives: $2C \dfrac{d}{dt} C = 0 + 2B \dfrac{d}{dt} B$

Rearranging gives: $\dfrac{d}{dt} C = \dfrac{2B}{2C} \dfrac{d}{dt} B$

When B is 5 and (dB/dt) is 2, then C can be found at B = 5 using $A^2 + B^2 = C^2$. Therefore:

$$C = (5^2 + 5^2)^{1/2} = (25 + 25)^{1/2} = (50)^{1/2} = 5(2)^{1/2}$$

Differentiating:

$$\frac{dC}{dt} = \frac{B}{C} \frac{dB}{dt} = \frac{5}{5\sqrt{2}} (2) = 2/(2)^{1/2} = \frac{\sqrt{2}}{\sqrt{2}} \frac{2}{\sqrt{2}} = \frac{2}{2}\sqrt{2} = \sqrt{2}$$

Therefore, when B is at 5 cm, $dC/dt = \sqrt{2}$ cm/s.

2.21 Differentiating Trigonometric Functions

• This section includes the relationship between sine and cosine with respect to their first and second derivatives, the derivatives of tangent, cotangent, secant, and cosecant.

Sine and Cosine

• The *derivative of sine* is cosine and the *derivative of cosine* is −sine. This can be visualized by comparing the slopes of sine and cosine curves at various points along their graphs.

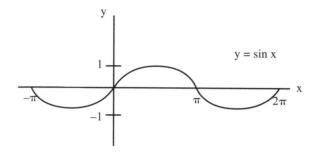

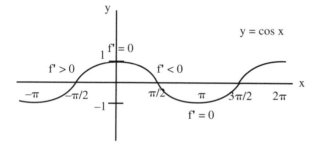

The derivative of the sine curve is zero at the top and bottom points on the curve where it is horizontal, $-\pi/2$, $\pi/2$, $(3/2)\pi$, $(5/2)\pi$, etc. The maximum rates of change of sine in the positive direction occur at points on the curve at 0, 2π, 4π, etc. The maximum rates of change of the sine curve in the negative direction occur at points $-\pi$, π, 3π, etc.

The points where the derivative of sine is zero correspond to points where the cosine curve crosses zero. The points where the derivative of sine is maximum-positive correspond to maximum points on the cosine curve. Similarly, the points where the derivative on the sine curve is maximum-negative correspond to the most negative points on the cosine curve.

The *cosine curve* is the *sine curve* shifted to the left by $(1/2)\pi$, which is consistent with the trigonometric identity, $\cos x = \sin(x + \pi/2)$

• The derivatives of sine and cosine can be verified using the definition of the derivative and the *addition formulas for sine and cosine*:

$$\sin(x+h) = \sin x \cos h + \cos x \sin h$$
$$\cos(x+h) = \cos x \cos h - \sin x \sin h$$

For $f(x) = \sin x$, find $f'(x)$:

$$f'(x) = \lim_{h \to 0} \frac{\sin(x+h) - \sin x}{h}$$

$$= \lim_{h \to 0} \frac{\sin x \, \cos h + \cos x \, \sin h - \sin x}{h}$$

$$= \lim_{h \to 0} \left(\frac{\sin x \, \cos h - \sin x}{h} + \frac{\cos x \, \sin h}{h} \right)$$

$$= \lim_{h \to 0} \frac{\sin x \, (\cos h - 1)}{h} + \lim_{h \to 0} \frac{\cos x \, \sin h}{h}$$

$$= \sin x \, \lim_{h \to 0} \frac{\cos h - 1}{h} + \cos x \, \lim_{h \to 0} \frac{\sin h}{h}$$

As $h \to 0$, $\dfrac{\cos h - 1}{h} \to \dfrac{\cos 0.000001 - 1}{0.000001} = \dfrac{1 - 1}{0.000001} = 0$

As $h \to 0$, $\dfrac{\sin h}{h} \to \dfrac{\sin 0.000001}{0.000001} = \dfrac{0.000001}{0.000001} = 1$

Therefore, $f'(x) = 0 + \cos x = \cos x$.

Similarly, if $f(x) = \cos x$, find $f'(x)$:

$$f'(x) = \lim_{h \to 0} \frac{\cos(x+h) - \cos x}{h}$$

$$= \lim_{h \to 0} \frac{\cos x \, \cos h - \sin x \, \sin h - \cos x}{h}$$

$$= \lim_{h \to 0} \frac{\cos x \, \cos h - \cos x - \sin x \, \sin h}{h}$$

$$= \lim_{h \to 0} \frac{\cos x \, (\cos h - 1)}{h} - \lim_{h \to 0} \frac{\sin x \, \sin h}{h}$$

$$= \cos x \, \lim_{h \to 0} \frac{\cos h - 1}{h} - \sin x \, \lim_{h \to 0} \frac{\sin h}{h}$$

Therefore, $f'(x) = 0 - \sin x = -\sin x$.

• In general, the *second derivative of a function* indicates whether the curve is *concave up* (*positive derivative*) or *concave down* (*negative derivative*) at the point on the curve where the derivative is taken. This can be observed on the following graphs of sine and cosine.

For example at x = 0, sine begins moving up and its derivative (cosine) is positive. However, the curve is moving into a concave-down shape, which is consistent with the second derivative becoming negative.

Similarly at x = 0, cosine begins and slopes downward, its first derivative (sine) becomes negative. However, the second derivative is also negative, which is consistent with the curve moving along a concave-down shape.

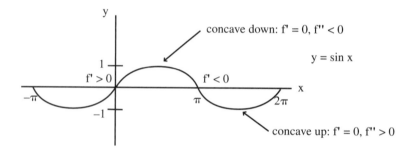

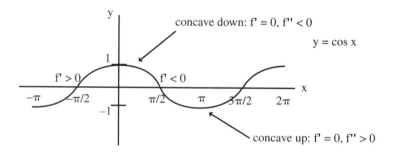

The *slope* at each point on the sine curve is given by the value of the cosine curve at that point.

• *Second derivatives of sine and cosine* are:

$$\frac{d^2}{dx^2}\sin x = \frac{d}{dx}\cos x = -\sin x$$

$$\frac{d^2}{dx^2}\cos x = -\frac{d}{dx}\sin x = -\cos x$$

Tangent, Cotangent, Secant, and Cosecant

• The *derivative of tangent* can be easily determined using the fact that tan x = (sin x / cos x) and the rule for differentiating quotients (described in Section 2.18).

The *quotient rule* is $\dfrac{d}{dx}\dfrac{f(x)}{g(x)} = \dfrac{f'(x)g(x) - f(x)g'(x)}{g(x)^2}$

in this case, f(x) = sin x and g(x) = cos x:

$$\frac{d}{dx}\tan x = \frac{d}{dx}\frac{\sin x}{\cos x} = \frac{((d/dx)\sin x)(\cos x) - (\sin x)((d/dx)\cos x)}{(\cos x)^2}$$

$$= \frac{\cos x \cos x - \sin x\,(-\sin x)}{(\cos x)^2} = \frac{\cos^2 x + \sin^2 x}{\cos^2 x} = \frac{1}{\cos^2 x}$$

(Remember, cos²x + sin²x = 1.)

• The *derivatives of cotangent, secant, and cosecant* are:

(d/dx)cot x = −csc²x
(d/dx)sec x = sec x tan x
(d/dx)csc x = −csc x cot x

2.22 Inverse Functions and Inverse Trigonometric Functions and Their Derivatives

• This section includes a brief summary of inverse functions, the derivative of inverse functions, inverse trigonometric functions, and their derivatives.

• *Inverse functions* are functions that result in the same value of x after the operations of the two functions are performed. In inverse functions, the operations of each function are the reverse of the other function. If f is the inverse of g then g is the inverse of f. *Notation for the inverse of function f is f⁻¹*.

• An inverse of a function has its domain and range equal to the range and domain, respectively, of the original function. If f(x) = y, then f⁻¹(y) = x. For a function f(x,y) that has only one y value for each x value, then there exists an inverse function represented by f⁻¹(y,x). A function has an inverse if its graph intersects any horizontal line no more than once.

• If function f is represented by $f(x) = u$, then its inverse f^{-1} can be found by solving $f(x) = u$ for x in terms of u: $f^{-1}u = f^{-1}(f(x)) = x$. Therefore, if $f(x) = u$ then $f^{-1}(u) = x$, or if $f^{-1}(u) = x$ then $f(x) = u$. For more complicated or *composite functions*, if $y = f[u(x)]$, then the inverse can be written in the opposite order: $x = u^{-1}(f^{-1}(y))$

• Not all functions have inverses. If a function has more than one solution, it does not have an inverse. If $u(x) = z$, only one x can result, $x = u^{-1}(z)$. If there is more than one solution for $u^{-1}(z)$, it will not be the inverse of $u(x) = z$.

• When functions f and u are inverse functions, then they will return to the first value. For example, if $y = f(x) = 2x - 1$ and $x = f^{-1}(y) = (y + 1)/2$ are inverses, and if $x = 3$, then by substituting for x:

   $f(3) = 2(3) - 1 = 5$

By substituting 5 into inverse function:

   $f^{-1}(5) = (5 + 1)/2 = 3$

which results in the starting point.

• *Graphs of inverse functions* are mirror images across the $x = z$ line. For example, if $z = u(x) = 2x$, then $x = (1/2)z$. The slopes are $(dz/dx) = 2$ and $(dx/dz) = 1/2$.

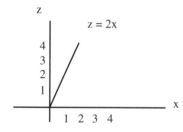
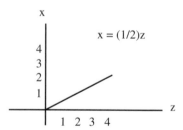

• Following are examples of functions and their inverses:

   $z = x^2$ is the inverse of: $x = \sqrt{z}$ or $x = z^{1/2}$
   $z = e^x$ is the inverse of: $x = \ln z$
   $z = a^x$ is the inverse of: $x = \log_a z$

• The *derivatives of inverse functions* $y = f(x)$ and $x = f^{-1}(y)$, have the property: $(dy/dx)(dx/dy) = 1$.

For example, using inverse functions:

$y = f(x) = 2x - 1$ and $x = f^{-1}(y) = (y + 1)/2$
$dy/dx = 2$ and $dx/dy = 1/2$

Therefore, $(dy/dx)(dx/dy) = (2)(1/2) = 1$.

• If $f[u(x)] = x$ is an inverse function, then applying the *chain rule* gives:
$f'[u(x)] \times u'(x) = 1$.

If $y = u(x)$ and $x = f(y)$, then the rule is written:

$(dx/dy)(dy/dx) = 1$, where the slope of $y = u(x)$
multiplied by the slope of $x = u^{-1}(y)$ is equal to 1.

• *Inverses of trigonometric functions* introduced in Chapter 1 exist in defined intervals. For example, the *inverse of sine* is $\sin^{-1}y = x$ for $1 \geq y \geq -1$, which pertains to $\sin x = y$ for $\pi/2 \geq x \geq -\pi/2$. The inverse brings y back to x.

The graph of $y = \sin x$ is a mirror image of $\sin^{-1}y = x$.

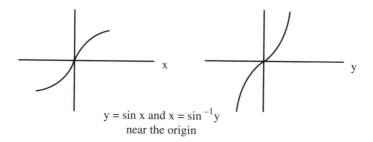

$y = \sin x$ and $x = \sin^{-1}y$
near the origin

Only certain intervals of the sine function have inverses:

In the interval $\pi/2 \geq x \geq -\pi/2$, $\sin^{-1}(\sin x) = x$.
In the interval $1 \geq y \geq -1$, $\sin(\sin^{-1}y) = y$.

There are many points on the sine function where $\sin x = 0$.

• The *derivative of inverse sine*, where $x = \sin^{-1}y$, exists in the interval $1 \geq y \geq -1$ and $\pi/2 \geq x \geq -\pi/2$. The derivative of the inverse function equals one over the derivative of the original function. The derivative of the inverse function $x = \sin^{-1}y$ can be found using the derivative of the original function $y = \sin x$, where:

$(dy/dx) = \cos x$

then by rearranging,

$(dx/dy) = 1/\cos x$

Going further and using the trigonometric identity $\cos^2x + \sin^2x = 1$, or equivalently, $\cos x = (1 - \sin^2x)^{1/2}$, combined with $\sin x = y$ and squaring $\sin^2x = y^2$:

$(dx/dy) = (d/dy)\sin^{-1}y = 1/\cos x = 1/(1 - y^2)^{1/2}$

• The *derivative of inverse cosine*, where $x = \cos^{-1}y$, exists in the interval $1 \geq y \geq -1$ and $\pi \geq x \geq 0$. The derivative of the inverse function $x = \cos^{-1}y$ can be found using the derivative of the original function $y = \cos x$, where:

$(dy/dx) = -\sin x$

then by rearranging,

$(dx/dy) = -1/\sin x$

Going further and using the trigonometric identity $\cos^2x + \sin^2x = 1$, or equivalently, $\sin x = (1 - \cos^2x)^{1/2}$, combined with $\cos x = y$ and squaring $\cos^2x = y^2$:

$(dx/dy) = (d/dy)\cos^{-1}y = -1/\sin x = -1/(1 - y^2)^{1/2}$

• For the tangent function $\tan x = y$, the inverse, $\tan^{-1}y = x$ exists for $\pi/2 > x > -\pi/2$. The *derivative of inverse tangent*, where $x = \tan^{-1}y$, exists for $\infty > y > -\infty$ and $\pi/2 > x > -\pi/2$. The derivative of the inverse function $x = \tan^{-1}y$ can be found using the derivative of the original function $y = \tan x$, where:

$(dy/dx) = -\sec^2x$

then by rearranging,

$(dx/dy) = -1/\sec^2x$

Going further and using the trigonometric identity $\sec^2 x = 1 + \tan^2 x$, combined with $\tan x = y$ and squaring $\tan^2 x = y^2$:

$$(dx/dy) = (d/dy)\tan^{-1}y = -1/\sec^2 x = 1/(1 + \tan^2 x) = 1/(1 + y^2)$$

• The following are *derivatives of inverse cotangent, secant, and cosecant*:

Derivative of $\cot^{-1}y$ is $-1/(1 + y^2)$,

  for $\infty > y > -\infty$ and $\pi > x > 0$

Derivative of $\sec^{-1}y$ is $1/|y|(y^2 - 1)^{1/2}$,

  for $1 > y > -1$ and $\pi > x > 0$

Derivative of $\csc^{-1}y$ is $-1/|y|(y^2 - 1)^{1/2}$,

  for $1 > y > -1$ and $\pi/2 > x > -\pi/2$

2.23 Differentiating Hyperbolic Functions

• This section includes differentiating hyperbolic functions and their inverses.

• *Hyperbolic functions* include sinh, cosh, tanh, coth, sech, and csch, and are introduced in Chapter 1. The properties of sine and cosine are reflected in the hyperbolic sine and cosine, sinh and cosh, respectively, such as:

$$(\cosh x)^2 - (\sinh x)^2 = 1$$

which is similar to:

$$(\cos x)^2 + (\sin x)^2 = 1$$

• *Derivatives of sinh and cosh* are similar to the derivatives of sine and cosine.

$$(d/dx)\sinh x = \cosh x$$

which is similar to:

$$(d/dx)\sin x = \cos x$$

$$(d/dx)\cosh x = \sinh x$$

which is similar to:

   $(d/dx)\cos x = -\sin x$ (without the "−" sign)

• The *derivatives of tanh, coth, sech, and csch* are:

   $(d/dx)\tanh x = \text{sech}^2 x$

   $(d/dx)\coth x = -\text{csch}^2 x$

   $(d/dx)\text{sech } x = -\text{sech } x \tanh x$

   $(d/dx)\text{csch } x = -\text{csch } x \coth x$

• *Derivatives of inverse hyperbolic functions $\sinh^{-1}x$, $\tanh^{-1}x$, and $\text{sech}^{-1}x$ are:*

The inverse of $x = \sinh y$ is $y = \sinh^{-1}x$

   $(d/dx)\sinh^{-1}x = 1/(1 + x^2)^{1/2}$

which is similar to:

   $(d/dx)\sin^{-1}x = 1/(1 - x^2)^{1/2}$

The inverse of $x = \tanh y$ is $y = \tanh^{-1}x$

   $(d/dx)\tanh^{-1}x = 1/(1 - x^2)$

which is similar to:

   $(d/dx)\tan^{-1}x = 1/(1 + x^2)$

The inverse of $x = \text{sech } y$ is $y = \text{sech}^{-1}x$

   $(d/dx)\text{sech}^{-1}x = \pm 1/[x(1 - x^2)^{1/2}]$

which is similar to:

   $(d/dx)\sec^{-1}x = 1/[x(1 - x^2)^{1/2}]$

• The *derivatives of inverse hyperbolic functions $\cosh^{-1}x$, $\text{csch}^{-1}x$, and $\coth^{-1}x$ are:*

   $(d/dx)\cosh^{-1}x = 1/(x^2 - 1)^{1/2}$

   $(d/dx)\text{csch}^{-1}x = \pm 1/[x(1 + x^2)^{1/2}]$

   $(d/dx)\coth^{-1}x = 1/(1 - x^2)$

2.24 Differentiating Multivariable Functions

• This section introduces *differentiating simple multivariable* functions. See Chapter 6 for a complete discussion on differentiating multivariable functions.

• When a function that contains more than one variable is differentiated, the derivative formula can be applied to the variable that is being differentiated "with respect to," while the other variable(s) is held constant. The variable being differentiated "with respect to," is designated using d/dx, d/dy, d/dz, etc., for x, y, z, respectively.

• For example, differentiate the following simple functions with respect to x, y, and z:

$$\frac{d}{dx}(x^2y^2) = 2xy^2 \qquad \text{Differentiated with respect to x.}$$

$$\frac{d}{dy}(x^2y^2) = 2x^2y \qquad \text{Differentiated with respect to y.}$$

$$\frac{d}{dz}(x^2y^2z^2) = 2x^2y^2z \qquad \text{Differentiated with respect to z.}$$

2.25 Differentiation of Implicit Vs. Explicit Functions

• This section provides a brief explanation of implicit differentiation including two examples.

• If y is given explicitly as a function of x, it is not difficult to obtain (dy/dx) because if y = f(x), then (dy/dx) = f'(x). This is *explicit differentiation*. In explicit differentiation, y can be isolated and the equation can be solved for y, then differentiated. However, if y is given implicitly as a function of x, F(x,y) = 0, then the function can be differentiated as it is and then solved for dy/dx rather than attempting to isolate y first, which may not always be possible.

• For a function that cannot be solved for y first or is left in implicit form by choice, the equation can be differentiated implicitly as it is "term by term," then solved for (dy/dx) in terms of x and y. This is called *implicit differentiation*.

• **Example:** Evaluate (dy/dx) for $x^4 + x^2y^3 - y^6 + 4 = 0$, which implicitly gives y as a function of x.

Taking the derivative of each term gives:

$$4x^3 + 2xy^3 + 3x^2y^2 \frac{dy}{dx} - 6y^5 \frac{dy}{dx} = 0$$

Solving for (dy/dx):

$$3x^2y^2 \frac{dy}{dx} - 6y^5 \frac{dy}{dx} = -4x^3 - 2xy^3$$

$$\frac{dy}{dx} = [-4x^3 - 2xy^3] / [3x^2y^2 - 6y^5]$$

• **Example:** Evaluate (dy/dx) for $y = e^{yx} + \sin x$, which implicitly gives y as a function of x.

Taking the derivative of each term gives:

$$\frac{dy}{dx} = ye^{yx} + xe^{yx} \frac{dy}{dx} + \cos x$$

Solving for (dy/dx):

$$\frac{dy}{dx} - xe^{yx} \frac{dy}{dx} = ye^{yx} + \cos x$$

$$\frac{dy}{dx} = [ye^{yx} + \cos x] / [1 - xe^{yx}]$$

2.26 Selected Rules of Differentiation

• This section provides a summary of selected rules of differentiation. Note that in the following functions n is a positive integer and u and v are functions of x.

Function	Derivative
• $y = u^n$,	$\dfrac{dy}{dx} = nu^{n-1} \dfrac{du}{dx}$
• $y = a^u$,	$\dfrac{dy}{dx} = a^u \log a \dfrac{du}{dx}$
• $y = u^v$,	$\dfrac{dy}{dx} = vu^{v-1} \dfrac{du}{dx} + u^v \log u \dfrac{dv}{dx}$
• $y = e^u$,	$\dfrac{dy}{dx} = e^u \dfrac{du}{dx}$
• $y = \log u$,	$\dfrac{dy}{dx} = (1/u) \dfrac{du}{dx}$ providing $u \neq 0$.
• $y = \cos u$,	$\dfrac{dy}{dx} = -\sin u \dfrac{du}{dx}$
• $y = \sin u$,	$\dfrac{dy}{dx} = \cos u \dfrac{du}{dx}$

2.27 Minimum, Maximum, and the First and Second Derivatives

• This section includes local and global minimum and maximum points and the first and second derivatives.

• Evaluating first and second derivatives of functions to find *minimum* and *maximum* points is a common application of the derivative. When experiments or evaluations are conducted in science, business, engineering, etc., data is gathered, relationships are developed, and graphs are constructed in order to assist in the understanding of the data and to predict future patterns and events. Information depicted in graphs, such as where the graph is rising or falling, convex or concave, or where the high and low points are located, correspond to maximum and minimum values and is crucial to the evaluation of the data.

• Consider the graph of a continuous function f:

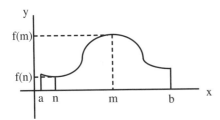

If the highest point on f is the point (m,f(m)), then f(m) is the maximum value of f and f(m) ≥ f(x) for all x. In this graph there are two extrema points in between a and b, a minimum and a maximum where the derivative is zero.

To find the *global* or *local extrema* of a function, the graph can be inspected or the derivative can be evaluated. At the extrema points, the derivative of a function f is equal to zero. In this graph if f(n) is a minimum point and f(m) is a maximum point and if f'(n) and f'(m) exist, then f'(n) = 0 and f'(m) = 0.

• The graph of a function has a minimum or maximum point where the *slope is zero* and therefore the *derivative is also zero*, f'(x) = 0. In a region of a graph of a function where the *graph is horizontal* the first derivative of the function is equal to zero. A point where the graph of a function is horizontal may represent a *minimum* or *maximum* point. A minimum or maximum on the graph may be the minimum or maximum of the function, or there may be many "local" minimum or maximum points called *local extrema*. There can only be one global minimum and one global maximum but there may be many local extrema points.

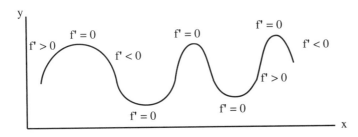

The *sign of the derivative* of a function describes the shape of the graph of the function at the point where the derivative is taken.

If f(x) is *decreasing* as x is increasing, the sign of the derivative is negative. Therefore, f'(x) < 0 where the graph of f is decreasing.

If f(x) is *increasing* as x is increasing, the sign of the derivative is positive. Therefore, f'(x) > 0 where the graph of f is increasing.

If the graph of the function is *horizontal*, the derivative of f is zero. Therefore, f'(x) = 0 where the graph of f is horizontal.

The sign of f'(x) changes from positive to negative or negative to positive as a maximum or minimum point is crossed.

• There are examples where the graph of a function will not have a minimum or maximum, such as if the graph forms a straight horizontal or vertical line. For example:

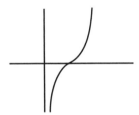

• As a general rule, for a given function f, all values of x where f'(x) = 0 or where f'(x) is undefined, represent all possible extrema. There may, however, be cases where f'(x) = 0 but an extrema does not exist.

• By taking the *second derivative* of a function where the first derivative is zero, it can be determined whether the graph of that function is at a mini-mum and, therefore, *concave up* or at a maximum and, therefore, *concave down*. The second derivative provides information about change in slope, or the rate of change of what is changing, such as the growth rate of a population.

• If some point P is in the domain set of function f and if f'(P) exists, then the second derivative can be used to evaluate the shape of the graph as follows:

If f'(P) = 0 and if f''(P) > 0, the graph of function f is *concave up* at P and f has a *minimum* at P. Also, the slope of the curve or tangent lines drawn to the curve will begin to increase.

If f'(P) = 0 and if f''(P) < 0, the graph of function f is *concave down* at P and f has a *maximum* at P. Also, the slope of the curve or tangent lines drawn to the curve will begin to decrease.

In other words, if f'(P) exists, and if f'(P) = 0:

    If f''(P) > 0 then f has a minimum at P, or
    if f''(P) < 0 then f has a maximum at P.

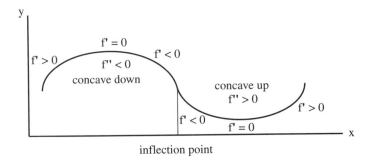

inflection point

Note that in the above graph an *inflection point* occurs where the tangent line crosses the curve. Also, an inflection point occurs where f''(x) changes from positive to negative or negative to positive, and where the curve is concave up on one side and concave down on the other side.

• If the *second derivative* of a function is zero, then it does not provide information regarding whether the function is at a maximum or minimum. In this situation, information can be obtained in the region where f'(x) = 0 such that if f'(x) changes from positive to negative at f'(x) = 0, then there is a maximum at that point. Conversely, if f'(x) changes from negative to positive at f'(x) = 0, then there is a minimum at that point.

• To solve problems where minimum or maximum values need to be found, first describe the problem in terms of a function or equation, then determine f'(x) and solve for f'(x) = 0. To locate all possible extrema (minimum or maximum values of f) within some interval between x = a and x = b or between points (a,f(a)) and (b,f(b)):

(a.) Find all x values that satisfy f'(x) = 0 or f'(x) = undefined.

(b.) Evaluate each x value found in the first step by substituting it into the function f.

(c.) Evaluate values of x at the ends of the interval (at a and b) to find f(a) and f(b).

(d.) The largest value in the second step is the maximum of f(x) and the smallest value is the minimum of f(x) within the interval a-b.

(e.) Identify whether the extrema represent a minimum or maximum by determining f''(x).

• **Example:** Find the minimum and maximum of function $f(x) = x^2 + 2x$ between the interval of x = 0 and x = −2 where $-2 \le x \le 0$.

First find all x values that satisfy f'(x) = 0 or f'(x) = undefined. Differentiate:

$$f'(x) = (d/dx)x^2 + (d/dx)2x = 2x + 2$$

Where does f'(x) = 0?

Because f'(x) = 2x + 2, set 2x + 2 = 0:

$$2x + 2 = 0$$

Solve for x:

$$2x = -2$$
$$x = -2/2 = -1$$

Evaluate each x value found by substituting it into the function f. Evaluate f(x) at x = −1.

$$f(-1) = x^2 + 2x = (-1)^2 + 2(-1) = 1 + -2 = -1$$

Evaluate the values of x at the ends of the interval (at a and b) to find f(a) and f(b).

Evaluate $f(x) = x^2 + 2x$ at the end points -2 and 0.

$$f(-2) = (-2)^2 + 2(-2) = 4 + -4 = 0$$
$$f(0) = (0)^2 + 2(0) = 0 + 0 = 0$$

Therefore, the number for the critical points of f over this interval $(-2 \leq x \leq 0)$ are: $f(-1) = -1$, $f(0) = 0$ and $f(-2) = 0$.

The largest and smallest values from the second step are the maximum of $f(x)$ and the minimum of $f(x)$ within the interval a-b. In this example, only $f(-1) = -1$ was derived from the second step. The largest and smallest numbers computed overall are 0 and -1, which represent the minimum and maximum points.

Plot the function $f(x) = x^2 + 2x$ between the interval of $x = 0$ and $x = -2$. Select x values at and near the minimum and maximum points, and solve for $f(x)$.

Values for x are $-3, -2, -1, 0, 1$, resulting in $f(x)$ values $3, 0, -1, 0, 3$.

Resulting pairs are $(-3,3), (-2,0), (-1,-1), (0,0), (1,3)$.

Graphing the pairs is depicted as:

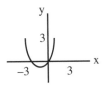

Therefore, $f'(-1) = 0$ within the interval a-b from $x = -2$ to $x = 0$, the graph of the function $f(x) = x^2 + 2x$ depicts a minimum at $f(-1)$, and crosses zero at $x = -2$ and $x = 0$.

Evaluating the second derivative of this function, $f(x) = x^2 + 2x$, will determine whether there is a minimum or a maximum at the point where $f'(x) = 0$ (and therefore verify the result of the graph).

$$f'(x) = 2x + 2$$

Taking the second derivative:

$$f''(x) = (d/dx)2x + (d/dx)2 = 2 + 0 = 2$$

Using the second derivative rule, because 2 is a positive number, the graph of f(x) at x = −1 is concave up and is at a minimum. This was depicted in the graph.

2.28 Notes on Local Linearity, Approximating Slope of Curve, and Numerical Methods

• This section includes a brief introduction of local linearity and the tangent line approximation and a brief explanation of Newton's method for equations in the form f(x) = 0.

• When *calculating approximate values for complicated functions*, it is sometimes possible to focus in on a small region of the graph of a function, and look at that region as if it were linear. This is sometimes referred to as a point of *local linearity*. In the region of a point on the graph of a function, a tangent line can be drawn and the slope of the tangent line is the derivative of the function at that point. The *equation for a tangent* line passing through point (a,f(a)) is: y − f(a) = f'(a)(x − a).

• The *equation for the tangent line* y − f(a) = f'(a)(x − a), at y = f(x) and x = a can be derived using:

$$f'(a) \approx [f(a + h) − f(a)]/h$$

By rearranging:

$$f'(a)(h) \approx f(a + h) − f(a)$$
$$f(a + h) \approx f'(a)(h) + f(a)$$

Substituting x − a = h and x = a + h:

$$f(a + h) \approx f'(a)(x − a) + f(a)$$
$$f(x) \approx f'(a)(x − a) + f(a)$$

Substituting y = f(x):

$$y − f(a) = f'(a)(x − a)$$

This equation can be used to *linearize* a region of f near x = a for the tangent through (a, f(a)) for curved functions.

• For example, the *tangent line approximation* for f(x) = cos x, where x is near a = 0 can be calculated using:

$$f(x) \approx f'(0)(x - 0) + f(0)$$

Substitute in for each term:

Left side term:

$$f(x) = \cos x$$

The first term:

$$f'(0) = -\sin 0 = 0$$
$$\text{and } x - 0 = x$$

The second term:

$$f(0) = \cos 0 = 1$$

Then the equation $f(x) \approx f'(0)(x - 0) + f(0)$ becomes:

$$\cos x \approx (0)(x) + 1 = 1$$
$$\cos x \approx 1$$

Therefore, the tangent line approximation for

$$f(x) \text{ near } x = 0 \text{ is } y = f(x) = 1.$$

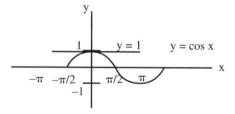

• Similarly for f(x) = sin x near x = 0, the *tangent line approximation* can be calculated using:

$$f(x) \approx f'(0)(x - 0) + f(0)$$

Substitute in for each term:

Left side term:

$$f(x) = \sin x$$

The first term:

$$f'(0) = \cos 0 = 1$$
and $x - 0 = x$

The second term:

$$f(0) = \sin 0 = 0$$

Then the equation $f(x) \approx f'(0)(x - 0) + f(0)$ becomes:

$$\sin x \approx (1)(x) + 0 = x$$
$$\sin x \approx x$$

Therefore, the tangent line approximation for

$f(x)$ near $x = 0$ is $y = f(x) = x$.

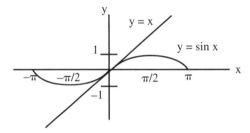

• The tangent line drawn on selected points of the graph of a function can be used in *numerical differentiation methods*, such as *Newton's method*. *Numerical methods* are sometimes required to estimate and solve equations, such as finding the roots of a high-degree polynomial. In general, numerical methods can be applied to programming when a function is translated into an algorithm, solving systems of linear equations, and numerical solutions to ordinary and partial differential equations.

• The Newton method can be applied to solve equations in the form $f(x) = 0$ where $f'(x)$ exists and is continuous.

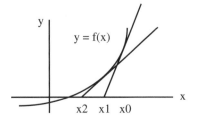

In this method, the graph of f is approximated using *tangent lines*, thereby determining the *roots* (x values) of f(x). First, a value for x_0 is selected from the graph of f, then a tangent is drawn at x_0, where x_1 is the intersection of the X-axis by the tangent to the curve at x_0. This process can be repeated beginning with the value of x_1 and drawing the tangent that intercepts the X-axis at x_2. Then, repeat for x_2 to get x_3 and so on until the x's converge.

Alternatively, after the first step the equation for a tangent line at f(x) = y, x = x_0 can be used:

$$y - f(x_0) = f'(x_0)(x - x_0)$$

where the tangent crosses the X-axis at y = 0, x = x_1, therefore:

$$0 - f(x_0) = f'(x_0)(x_1 - x_0)$$

Rearranging:

$$x_1 = x_0 - \frac{f(x_0)}{f'(x_0)}$$

This formula can be used repeatedly for x_2, x_3,...,x_{n+1}:

$$x_2 = x_1 - \frac{f(x_1)}{f'(x_1)}$$

$$x_{n+1} = x_n - \frac{f(x_n)}{f'(x_n)}$$

Using this formula, x_n should converge to a solution (root) of x if a solution exists. In cases where f(x) = 0 has no root or multiple roots, this formula will not converge to a single x value. Such cases include f(x) = $1 + x^2$.

Chapter

3

The Integral

3.1 Introduction

• *Integration* can be thought of as a sum of an infinite number of objects or sections that are infinitesimally small. The *integral* can be used to calculate area under a curve, area of a region or surface, volume of an object, average value of a function, work done, pressure, as well as the change in a function when its rate of change is known. The last example uses the Fundamental Theorem of Calculus.

• Just as the derivative can be thought of as the limit of differences, the integral can be thought of as the limit of sums. On the graph of a function, the derivative can be represented by slope of the curve and the integral can be represented by area under the curve. The derivative of distance x is velocity v, and the area under the curve of v is x. The integral of velocity is distance and the integral of acceleration is velocity. The integral can be found by constructing sums or by calculating the antiderivative or definite integral using formulas, techniques, and tables.

3.2 Sums and Sigma Notation

• This section includes a brief review of sums, sigma notation, calculating sums, properties of sums, and changing limits.

• *Sums* are used in the estimation of integrals pertaining to area and volume. Because sums are used to represent integrals, the following brief review of sums and sigma notation is included.

• The Greek letter *sigma* Σ is used to describe a *sequence* of numbers that are combined in a sum. The following are examples of sums:

$$1 + 2 + 3 + 4 = \sum_{i=1}^{4} i$$

$$1 + 2 + 3 + \ldots + n = \sum_{j=1}^{n} j$$

$$1/1 + 1/2 + 1/3 = \sum_{k=1}^{3} 1/k$$

$$1^3 + 2^3 + 3^3 = \sum_{i=1}^{3} i^3$$

$$a_1 + a_2 + a_3 + \ldots + a_{25} = \sum_{n=1}^{25} a_n$$

$$a_{15} + a_{16} + a_{17} + \ldots + a_{25} = \sum_{n=15}^{25} a_n$$

Sums do not have to begin at 1.

$$\sum_{j=3}^{5} j^2 = 3^2 + 4^2 + 5^2 = 9 + 16 + 25 = 50$$

• The *limits of the sum* indicate the first and last numbers in the sum. For example:

The limits for $\displaystyle\sum_{i=1}^{3} i^3$ are $i = 1$ to $i = 3$.

The letters i, j, and k are called *dummy variables* and represent the numbers in the sequence.

- Following are two examples of calculating sums:

(a.) $\sum_{k=2}^{5} 3k - 3$

where the first and last numbers to be substituted are 2 and 5, respectively.

$$3(2) - 3 = 6 - 3 = 3$$
$$3(3) - 3 = 9 - 3 = 6$$
$$3(4) - 3 = 12 - 3 = 9$$
$$3(5) - 3 = 15 - 3 = 12$$

Therefore, $\sum_{k=2}^{5} 3k - 3 = 3 + 6 + 9 + 12 = 30$.

(b.) $\sum_{k=1}^{4} 2k = 2(1) + 2(2) + 2(3) + 2(4) = 2 + 4 + 6 + 8 = 20$

- *Properties of sums* include:

(a.) If A_n and B_n each represent a sequence of numbers, then their sum can be written as:

$$\sum_{j=1}^{N} (A_j + B_j) = \sum_{j=1}^{N} A_j + \sum_{j=1}^{N} B_j$$

This property can be demonstrated as follows:

$$\sum_{j=1}^{N} (A_j + B_j)$$

$$= (A_1 + B_1) + (A_2 + B_2) + (A_3 + B_3) + ... + (A_N + B_N)$$
$$= (A_1 + A_2 + A_3 + ... + A_N) + (B_1 + B_2 + B_3 + ... + B_N)$$

$$= \sum_{j=1}^{N} A_j + \sum_{j=1}^{N} B_j$$

(b.) If A_n represents a sequence and c represents a number that is multiplied with the sequence, then:

$$\sum_{j=1}^{N} cA_j = c \sum_{j=1}^{N} A_j$$

This property can be demonstrated as follows:

$$\sum_{j=1}^{N} cA_j = (cA_1) + (cA_2) + (cA_3) + ... + (cA_N)$$

$$= c(A_1 + A_2 + A_3 + ... + A_N) = c \sum_{j=1}^{N} A_j$$

(c.) The sum of the first n terms in a sequence k can be calculated using the formula $(n/2)(n + 1)$:

$$\sum_{k=1}^{n} k = 1 + 2 + 3 + ... + (n - 1) + n = (n/2)(n + 1)$$

This formula can be demonstrated using the following two sums:

$$\sum_{k=1}^{4} k = 1 + 2 + 3 + 4 = 10$$

$$\sum_{k=1}^{4} k = (n/2)(n + 1) = (4/2)(4 + 1) = 2(5) = 10$$

To find a solution that does not begin with 1 consider the sum:

$$\sum_{k=6}^{10} k = 6 + 7 + 8 + 9 + 10 = 40$$

where $\sum_{k=6}^{10} k$ is equivalent to $\sum_{k=1}^{10} k - \sum_{k=1}^{5} k$

To calculate, subtract the formula for each sum:

$$[(10/2)(10 + 1)] - [(5/2)(5 + 1)] = 55 - 15 = 40$$

• Sometimes it is advantageous to *change the variables* and *limits of a sum*. This is possible as long as the changed sum is identical to the original sum. For example, the following notation represents the same sum:

$$\sum_{i=0}^{3} 3i = \sum_{j=1}^{4} 3(j-1), \text{ where } i = j - 1 \text{ or } j = i + 1.$$

This can be demonstrated as follows:

$$\sum_{i=0}^{3} 3i = 3(0) + 3(1) + 3(2) + 3(3) = 3 + 6 + 9 = 18$$

$$\sum_{j=1}^{4} 3(j-1) = 3(1 - 1) + 3(2 - 1) + 3(3 - 1) + 3(4 - 1)$$

$$= 3 + 6 + 9 = 18$$

Because i and j are *dummy variables*, it is okay to use either i or j after substitution. However, it is important to perform a change of limits appropriately. Therefore:

$$\sum_{i=1}^{4} 3(i-1) = \sum_{i=0}^{3} 3i$$

- *Products* are represented by the capital Greek letter *Pi or* Π and are similar to sums except that once all the terms have been established by substituting the integers consecutively (from the lower integer to the upper integer), the terms are multiplied with each other rather than added.

3.3 The Antiderivative or Indefinite Integral and the Integral Formula

- This section includes the definition and notation for the antiderivative or indefinite integral, the integral formula, the constant of integration, families of antiderivatives, the indefinite integral of acceleration and velocity, and the indefinite integral of a constant alone.

- The *antiderivative or indefinite integral* is approximately equal to the reverse of the derivative. The antiderivative or indefinite integral of a function f(x) is written:

$$\int f(x)\, dx$$

where $\int$ is the *integral symbol* and f(x) is the *integrand*.

- If the derivative of the function f(x) is the function F(x) so that df(x)/dx = F(x), then the antiderivative of F(x) is f(x) plus a *constant*.

$$\int F(x)\, dx = f(x) + c$$

where c represents an arbitrary *constant of integration* and dx indicates that *integration occurs with respect to* x.

- Remember the *derivative formula*:

$$dx^n/dx = nx^{n-1}$$

Similarly, there is an *integral formula* for calculating *antiderivatives or indefinite integrals*:

$$\int x^n \, dx = (1/(n+1))xn^{n+1} + c$$

where c represents a constant value and is called the *constant of integration*.

Note: The integral formula is an important formula and is used frequently in calculus.

- The derivative of the *antiderivative formula or integral formula* can be evaluated using the derivative formula:

$$\frac{d}{dx}\left[\frac{1}{n+1}x^{n+1} + c\right] = \frac{d}{dx}\left[\frac{1}{n+1}x^{n+1}\right] + \frac{dc}{dx}$$

$$= (n+1)\frac{1}{n+1}x^{n+1-1} + 0 = \frac{n+1}{n+1}x^{n+0} + 0 = x^n$$

- A *constant* is added to the indefinite integral because the derivative of a constant is zero. Also, because the derivative is the rate of change of some function, it seems likely that several different functions could have the same rate of change. For example, calculating the derivatives using the derivative formula of the following three functions results in the same rate of change:

$$(d/dx)(2x^2 + 3) = 4x$$
$$(d/dx)(5^{1/2} + 2x^2 + 2) = 4x$$
$$(d/dx)(2x^2 + \pi) = 4x$$

Then, take the indefinite integral of 4x using the integral formula to illustrate that each function is different even though they all have the same rate of change (derivative):

$$\int 4x \, dx = (1/(1+1))4x^{1+1} + c = (1/2)4x^2 + c = 2x^2 + c$$

Therefore, in the three functions c represents 3, $(5^{1/2} + 2)$, and π.

• This example demonstrates that an indefinite integral represents a *family of functions*, each with a different value for c. A function and its *family of antiderivatives* can be represented graphically by raising or lowering the curve by the constant values.

The graph of df(x)/dx can be used to plot the graph of its antiderivatives f(x) because df(x)/dx is the slope of the curve represented by f(x) at any point along the curve. When df(x)/dx is positive (above the X-axis), f(x) is increasing, and when df(x)/dx is negative (below the X-axis), f(x) is decreasing. Also, when df(x)/dx is increasing along the X-axis, f(x) is concave up, and when df(x)/dx is decreasing along the X-axis, f(x) is concave down. Finally, when df(x)/dx crosses zero, f(x) has a local minimum or maximum.

For the graph below represented by df/dx, there is a family of curves represented by the graph of f(x):

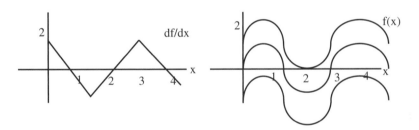

The slope of the antiderivative f(x) at any point, should be the value df/dx at that point. A *slope field* of f(x) can be constructed by drawing short lines at multiple points on its graph that represent the slope of the curve at each point.

• Remember that the derivative can represent the rate of change of distance x(t) as velocity v(t) and the rate of change of velocity v(t) as acceleration a(t), such that

$$dx(t)/dt = v(t) \text{ and } dv(t)/dt = a(t).$$

Conversely, the integral of *acceleration* is velocity and the integral of *velocity* is distance:

$$\int a(t) \, dt = v(t) + c$$

$$\int v(t) \, dt = x(t) + c$$

For example, using the integral formula,

$$\int x^n \, dx = (1/(n+1))x^{n+1} + c$$

acceleration, velocity, and distance can be represented as:

$$v = \int a \, dt = at + v_0$$

where v_0 represents a constant of integration.

Integrating again:

$$x = \int [at + v_0] \, dt = (1/2)at^2 + v_0t + x_0$$

where x_0 represents a constant of integration.

• The *integral of a constant* c_1 alone is equal to the constant c_1 multiplied by the variable the integral is being integrated with respect to (which is indicated by dx), plus another constant c_2:

$$\int c_1 \, dx = c_1x + c_2$$

3.4 The Definite Integral and the Fundamental Theorem of Calculus

• This section includes the definite integral, limits of integration, evaluating a definite integral using the Fundamental Theorem of Calculus, velocity and distance traveled, and calculating definite integrals using the integral formula and the Fundamental Theorem of Calculus.

• The indefinite integral represents a *family of functions* for different values of the constant of integration. Similarly, the *definite integral* represents a number pertaining to one of the functions of the indefinite integral such as the area under one of the curves. When the definite integral is evaluated, it is not necessary to add a *constant of integration*.

• If the endpoints of a function f(x) are set at specific values such as $x = a$ and $x = b$, where they may be depicted on the graph of f(x), then the function is integrated as a *definite integral*. The endpoint values for a definite integral are called the *limits of integration* and are shown at the ends of the integral symbol $_a\int^b$.

If the integral of the function f(x) between x = a and x = b is the function F(x), then F(x) must be evaluated at x = a and x = b. The symbol for *"evaluated at a and b"* is $|_a^b$ and it describes subtraction of the function at the top value b minus the function at the bottom value a:

$$F(x) \Big|_a^b = F(b) - F(a)$$

• The *Fundamental Theorem of Calculus* of a definite integral states that if f(x) is a continuous function between points x = a and x = b and f'(x) is the derivative of f(x), then:

$$\int_a^b f'(x)\, dx = f(b) - f(a)$$

Note: The Fundamental Theorem of Calculus is an important theorem and used frequently in calculus.

If F(x) is the antiderivative of f(x), so that:

$\int f(x) = F(x) + c$, or $F'(x) = f(x)$, then, the definite integral of f(x) between x = a and x = b is:

$$\int_a^b f(x)\, dx = F(b) - F(a)$$

• The *constant of integration* that results from an indefinite integral can be accounted for with respect to the definite integral as follows:

$$\int_a^b f(x)\, dx = [F(b) + c] - [F(a) + c] = F(b) - F(a)$$

• When the Fundamental Theorem is applied to a function defining *velocity* it provides a means to represent *distance traveled* in a designated time period by a definite integral. To analyze velocity and distance, the distance traveled can be represented by the definite integral of a velocity function. Because f'(t) = v(t), where f represents position, v represents velocity, and t represents time. The change in position or distance traveled from point a to point b can be written:

$$f(b) - f(a) = \int_a^b f'(t)\, dt$$

• The Fundamental Theorem is used when calculating many definite integrals. Applying the Fundamental Theorem is a shortcut to calculating the sums when the antiderivative can be found.

• The *integral formula for the definite integral* bounded by limits of integration x = a and x = b is used in conjunction with the Fundamental Theorem to calculate integrals:

$$\int_a^b x^n\, dx = (1/(n+1))(x^{n+1}) \Big|_a^b = (1/(n+1))(b^{n+1} - a^{n+1})$$

For simple integrals the integral formula is used to integrate the function and the Fundamental Theorem is used to evaluate the result. For more complex integrals, the integral formula must be combined with techniques such as integration by parts, substitution, and integral tables. These are discussed in the last four sections of this chapter.

• **Example:** Find the area of the function $f(x) = x^2$ between $x = 0$ and $x = 1$ by integrating (using the integral formula) and evaluating at the two x boundaries (or limits of integration).

$$\int_0^1 x^2 \, dx = (1/3) \, x^3 \big|_0^1 = (1/3)(1)^3 - (1/3)(0)^3 = 1/3$$

(See Section 3.6 "The integral and the area under a curve".)

3.5 Improper Integrals

• This section includes the definition of an improper integral, convergence of an improper integral, and applying the comparison test for convergence of an improper integral.

• A definite integral is called an *improper integral* if the *integrand is infinite* or becomes infinite between its limits, or if one or both of the *limits of integration are infinite*. An improper integral is discontinuous or diverges at one or more points in a function between its limits of integration.

• Even though some part or region of an improper integral is infinite, the integral may *converge* and the area under the curve may be finite. Consider the following integral with an infinite boundary that still converges to an area of 1:

$$\int_1^\infty x^{-2} \, dx = -x^{-1} \big|_1^\infty = [-1/\infty - -1/1] = 0 + 1 = 1$$

This can also be represented by:

$$\int_1^\infty x^{-2} \, dx = \lim_{b \to \infty} \int_1^b x^{-2} \, dx = \lim_{b \to \infty} [-x^{-1} \big|_1^b]$$

$$= \lim_{b \to \infty} [-1/b + 1/1] = 0 + 1 = 1$$

As b approaches infinity, 1/b approaches zero and the integral converges to 1.

In general, an integral in the form:

$$_1\!\int^\infty x^{-P}\, dx = {_1}\!\int^\infty 1/x^P\, dx \text{ where } P > 1$$

defines a finite area and therefore converges.

- An improper integral may not converge. For example:

$$\text{area under } {_1}\!\int^\infty x^{-1}\, dx = \ln x \,\big|_1^\infty = \infty$$

When an integral does not converge, it is said to *diverge*.

- An integral that approaches positive or negative infinity at either boundary or has an integrand that is infinite somewhere between the limits may still converge. This can be shown by *splitting* the integral into a sum of two integrals within the original limits. For example, if the integrand is *discontinuous* at some point c between the limits, the integral can be rewritten as the sum of two integrals each with one finite limit and with an infinite limit at point c. Then the integral can be written as:

$$_a\!\int^b f(x)\, dx = {_a}\!\int^c f(x)\, dx + {_c}\!\int^b f(x)\, dx$$

If both of the *integrals converge*, the sum of the resulting integrals will also converge.

Similarly, if both of the limits of integration of an integral are infinite, the integral can be rewritten as the sum of two integrals, each with a finite limit:

$$_{-\infty}\!\int^\infty f(x)\, dx = {_{-\infty}}\!\int^a f(x)\, dx + {_a}\!\int^\infty f(x)\, dx$$

where a is finite. Then if both of the integrals converge, the sum of the resulting integrals will also converge.

- The *Comparison Test* can be used to evaluate whether an improper integral will converge. This Comparison Test operates using the same concept as the Comparison Test described for convergence of a series in Chapter 4. The Comparison Test can be performed by finding another similar integral that fulfills the condition that if $0 \le f(x) \le g(x)$, then the area under $f(x)$ is less than the area under $g(x)$:

$$\text{If } \int g(x)\, dx < \infty, \text{ then } \int f(x)\, dx < \infty$$

In addition it follows that:

$$\text{if } \int f(x)\, dx = \infty, \text{ then } \int g(x)\, dx = \infty$$

To use this strategy to determine whether $f(x)$ converges, $f(x)$ must not be negative and must also be less than the function $g(x)$ that does converge.

• When the *Comparison Test* is applied to an infinite series with positive terms, the series is convergent if each term is less than or equal to each corresponding term in a series that is known to be convergent. Conversely, if each term in an unknown infinite series is greater than or equal to each corresponding term in a known divergent infinite series, then the unknown series is also divergent.

An example of a known convergent series that is used in the Comparison Test is the P Series:

$$1 + 1/2^P + 1/3^P + ... + 1/n^P + ...$$

This series converges when $P > 1$ and diverges when $P \leq 1$.

An example of an integral that is used as a comparison is:

$$_1\int^\infty 1/x^P \, dx$$

which converges when $P > 1$ and diverges when $P \leq 1$.

3.6 The Integral and the Area Under a Curve

• This section includes a theoretical explanation of the relationship between the area under a curve, distance, sums, and the definite integral.

• Integration provides a means to obtain *distance* information from *velocity* information. For example, if a bicyclist travels from the east side of town to the west side of town and if the velocity at each point in time is known, the distance traveled at each point can be determined. If information about time and velocity are known:

> time (hr): $t_0 = 0, t_1 = 1, t_2 = 2, t_3 = 3$
> velocity (mi/hr): $v_0 = 1, v_1 = 3, v_2 = 5, v_3 = 6$

Then the distance traveled $f(t)$ can be estimated using two slightly different approaches that give a lower and upper estimate for the actual value. The first calculates the velocity at the lower end of the time period and the second calculates the velocity at the end of each time period. This results in a lower and an upper estimate:

distance $f(t) = v(t_0)\Delta t + v(t_1)\Delta t + v(t_2)\Delta t$

distance $f(t) = v(t_1)\Delta t + v(t_2)\Delta t + v(t_3)\Delta t$

In this example the lower and upper estimates for distance $f(t)$ are:

$f(t) = (1 \text{ mi/h})(1 \text{ hr}) + (3 \text{ mi/hr})(1 \text{ hr}) + (5 \text{ mi/hr})(1 \text{ hr}) = 9 \text{ mi}$

$f(t) = (3 \text{ mi/h})(1 \text{ hr}) + (5 \text{ mi/hr})(1 \text{ hr}) + (6 \text{ mi/hr})(1 \text{ hr}) = 14 \text{ mi}$

The difference in the high and low estimates is $(14 - 9 = 5)$. If smaller time periods are chosen, such as 30 min., 5 min., 1 min., 1 sec., etc., then the difference between the upper and lower estimates becomes smaller and the overall estimate is better. The upper and lower estimates can be represented on a graph of velocity vs. time:

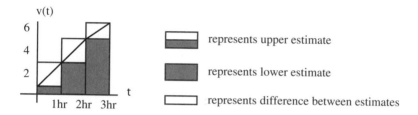

curve is drawn between upper and lower estimates

The *area* of each rectangle represents the distance traveled during each time period. The upper and lower rectangles of each period depicted represent the upper and lower estimates. The *sum* of the areas of the rectangles for all the time periods represents the *total distance traveled* (with the sum of the upper and lower rectangles representing the high and low estimates).

The difference between the estimates, as depicted on the graph, is the area of the rectangles that are between the upper and lower velocity values. The sum of the unshaded areas represents the total area of the differences. If the time period is reduced, the difference between the upper and lower estimates is proportionally reduced.

• The *exact value of distance traveled* in a given interval can be represented by taking the *limit of the sum of the areas* as the number of increments (rectangles) approaches infinity and, therefore, the rectangles become infinitesimally small. The estimates will converge to the accurate value of distance traveled as the number of increments measured approaches infinity. The limit of the sum as the number of increments in a defined region approaches infinity is the integral of the function defining the curve in that region.

In general, lower and upper *sums* (called *Riemann sums*) can be used to estimate an area. They both converge to the same integral and can be written in general terms as:

$$\lim_{n\to\infty} \sum_{i=0}^{n-1} f(x_i)\, \Delta x = {}_a\!\int^b f(x)\, dx$$

$$\lim_{n\to\infty} \sum_{i=1}^{n} f(x_i)\, \Delta x = {}_a\!\int^b f(x)\, dx$$

where a and b are the boundaries of a region that represents the area.

Note that the *limit as n→∞* is used with sums representing an integral so that infinite sums are not used. The *Riemann sum* defines the definite integral as the limit of the number of terms approaches infinity. There are examples, however, where the function is not continuous or forms an asymptote and the integral (and sum) will blow up and the integral cannot be used to define the area under some region of a curve between designated points.

• In general, the *integral is used to define the area under a curve on a graph of a function*. To use the integral to define the area under some region of a curve between points x = a and x = b, the curve in this region must be continuous and not extend into a vertical asymptote. Consider the following graph of function f(x):

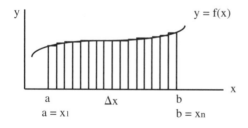

The striped pattern represents an approximation of the *area* under the curve of function y = f(x). The area under the curve between x = a and x = b is given by:

$${}_a\!\int^b f(x)\, dx$$

This is the *definite integral* of f(x) between x = a and x = b.

In the interval between x = a and x = b on the graph of f(x), the X-axis is divided into n equal parts of width Δx such that the Δx segments extend from the X-axis to the f(x) curve so that the area is divided into vertical rectangular strips. (Δ represents a small change in x.)

Between x = a and x = b, for the n rectangular strips, each strip is called the *ith strip*. The width of each strip is Δx and the height of each strip is y_i. The area of each strip is width times height, given by: $(y_i)(\Delta x)$ or $(f(x_i))(\Delta x)$

An *approximation for the total area* of f(x) between x = a and x = b is the sum of the areas of the n strips and is given by:

$$\text{Area} = \sum_{i=1}^{n} y_i \Delta x = y_1 \Delta x + y_2 \Delta x + y_3 \Delta x + ... + y_n \Delta x$$

Or equivalently:

$$\sum_{i=1}^{n} f(x_i) \Delta x = f(x_1)\Delta x + f(x_2)\Delta x + f(x_3)\Delta x + ... + f(x_n)\Delta x$$

If the width of Δx shrinks and the number of strips increases, the sum of the strips will represent a better approximation of the actual *area* under the curve. By taking the limit as the number of increments approaches infinity:

$$\text{Area} = \lim_{n \to \infty} \sum_{i=1}^{n} y_i \, \Delta x = {}_a\!\int^b f(x)\, dx$$

Note that Δx can be equivalently written (b–a)/n. Therefore, the area approximation can be written:

$$\text{Area} = \lim_{n \to \infty} [(b-a)/n] \sum_{i=1}^{n} y_i$$

In the graph of a function, as the number of increments increases in a specified interval and the rectangular area increments approach the area under the curve, then the sum approaches the integral given by: $\int f(x)\, dx$.

3.7 Estimating Integrals Using Sums and the Associated Error

• This section is an extension of the previous section and discusses the error associated with sums. It includes the midpoint rule, the trapezoid rule, and Simpson's rule.

• When *integrals are estimated* using upper and lower sums of rectangles, there is an error region above or below the curve. The total error for a curve bounded by two points is the sum of the error regions that lie between the upper and lower estimates. Lower and upper *Riemann Sums* (introduced in the previous section) used to estimate

$_a\int^b f(x)\ dx$ are:

$$f(x_0)\Delta x + f(x_1)\Delta x + f(x_2)\Delta x + \ldots + f(x_{n-1})\Delta x$$
$$f(x_1)\Delta x + f(x_2)\Delta x + f(x_3)\Delta x + \ldots + f(x_n)\Delta x$$

The error associated with the sums can be depicted as:

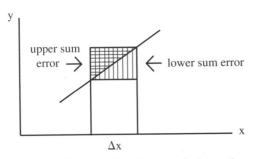

blowup of $y = f(x)$ depicting one Δx interval and error

The error resulting from summation can be reduced by increasing the number of increments (rectangles) and thus reducing their width.

• If integrals are estimated using sums of rectangles and the curve falls in the center (or midpoint) of each rectangle rather than the top or bottom, the error is reduced. Using this *midpoint rule*, the area estimated by the sum of rectangles (or increments) will be closer to the actual area under the curve. The *Riemann sum* using the midpoint rule is given by:

$$f(x_{1/2})\Delta x + f(x_{3/2})\Delta x + f(x_{5/2})\Delta x + \ldots + f(x_{n-1/2})\Delta x$$

The error for the midpoint rule can be depicted as:

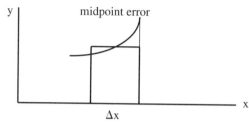

blowup of y = f(x) depicting one Δx interval
and error

- Integrals estimated using upper and lower sums of rectangles can have the error reduced by using an average of the two sums. This is referred to as the *trapezoid rule*. Also note that trapezoids fit under sloping lines. Using this trapezoid rule, the area is estimated by the sum of rectangles (or increments) where the average values of the measurements at the top and bottom of each rectangle are combined in the sum. The area for the rectangle from $f(x_0)$ to $f(x_1)$ is $(1/2)[f(x_0) + f(x_1)]\Delta x$. The Riemann sum using the trapezoid rule is given by:

$$(1/2)[f(x_0)\Delta x + f(x_1)\Delta x + f(x_2)\Delta x + ... + f(x_{n-1})\Delta x]$$
$$+ (1/2)[f(x_1)\Delta x + f(x_2)\Delta x + f(x_3)\Delta x + ... + f(x_n)\Delta x]$$
$$\text{or } \Delta x[(1/2)f(x_0) + f(x_1) + f(x_2) + ... + f(x_{n-1}) + (1/2)f(x_n)]$$

If the rectangles have different widths, the sum becomes:

$$(1/2)[f(x_0) + f(x_1)]\Delta x_1 + (1/2)[f(x_1) + f(x_2)]\Delta x_2$$
$$+ ... + (1/2)[f(x_{n-1}) + f(x_n)]\Delta x_n$$

where $\Delta x_i = (x_i - x_{i-1})$

The error for the trapezoid rule is depicted as:

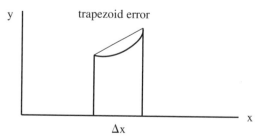

blowup of y = f(x) depicting one Δx interval
and error

- **Example:** Estimate $_0\int^2 x^2\,dx$ using the trapezoid rule with 4 subdivisions.

    Δx is equivalent to $(b-a)/n$ or $(2-0)/4 = 1/2$. Therefore,

    $\Delta x[(1/2)f(x_0) + f(x_1) + f(x_2) + ... + f(x_{n-1}) + (1/2)f(x_n)]$

becomes:

   $(1/2)[(1/2)(0)^2 + (1/2)^2 + 1^2 + (3/2)^2 + (1/2)(2)^2] =$
   $(1/2)(1/2)(0)^2 + (1/2)(1/2)^2 + (1/2)1^2 + (1/2)(3/2)^2 + (1/2)(1/2)(2)^2$
   $= 0 + 1/8 + 1/2 + 9/8 + 1 = 1/8 + 4/8 + 9/8 + 8/8 = 22/8 = 11/4$

Therefore, $_0\int^2 x^2\,dx$ is approximately equal to 11/4.

Comparing this result with calculating the definite integral directly:

   $_0\int^2 x^2\,dx = (1/3)x^3\,|_0^2 = (1/3)2^3 - 0 = 8/3$

Therefore, the error from using the trapezoid rule with

   $n = 4$ is $8/3 - 11/4 = 1/12$ or approximately 0.0833.

- Integrals can also be estimated using a method called *Simpson's rule*, which gives a better approximation than the trapezoid and midpoint rules. Simpson's rule combines the trapezoid and midpoint rules as a weighted average:

    $(1/3)[2 \times \text{(midpoint values)} + \text{(trapezoid values)}]$

The sum for Simpson's rule can be written:

   $(2/3)\Delta x[f(x_{1/2}) + f(x_{3/2}) + f(x_{5/2}) + ... + f(x_{n-1/2})]$
   $+ (1/3)\Delta x[(1/2)f(x_0) + f(x_1) + f(x_2) + ... + f(x_{n-1}) + (1/2)f(x_n)]$

- For example, use Simpson's rule to estimate $_0\int^2 x^2\,dx$, and compare it with the trapezoid rule above. Again, use 4 subdivisions.

Δx is equivalent to $(b-a)/n$ or $(2-0)/4 = 1/2$

Therefore, Simpson's rule is written:

   $(2/3)(1/2)[(1/4)^2 + (3/4)^2 + (5/4)^2 + (7/4)^2]$
   $+ (1/3)(1/2)[(1/2)(0)^2 + (1/2)^2 + 1^2 + (3/2)^2 + (1/2)(2)^2]$
   $= (2/3)(1/2)[(1/4)^2 + (3/4)^2 + (5/4)^2 + (7/4)^2] + (1/3)(11/4)$
   $= (1/3)[1/16 + 9/16 + 25/16 + 49/16] + 11/12$
   $= 21/12 + 11/12 = 32/12 = 8/3$

Therefore, $_0\int^2 x^2\, dx$ is approximately equal to 8/3.

This value of 8/3 is the exact value of the integral and therefore shows Simpson's rule to be an excellent estimate.

3.8 The Integral and the Average Value

• This section briefly discusses the relationship between the integral, the area under the curve, and the average value of a function in a defined interval.

• The integral not only represents the area under the curve on the graph of a function, but the integral also represents *average value* of a function in an interval.

The average value of f(x) in an interval a–b is the integral

$_a\int^b$ f(x) dx divided by the length of the a–b interval, given by (b − a):

average value = [1/(b − a)] $_a\int^b$ f(x) dx

The following figure depicts the average value where the integral represents the area under the curve, the average value of f is the height of the rectangle, and the width of the rectangle is (b − a).

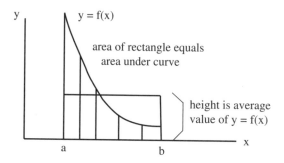

3.9 Area Below the X-axis, Even and Odd Functions, and Their Integrals

• This section includes a brief review of integrals and the areas above and below the X-axis, as well as even functions and odd functions and their integrals.

• In functions where the curve falls below the X-axis the *area* between the X-axis and the curve is negative in value and subtracts from the area above the X-axis.

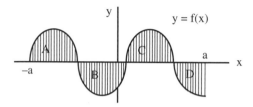

The graph of y = f(x) between x = −a and x = a is given by:

$$\int_{-a}^{a} f(x)\, dx = (\text{area A}) + (\text{area C}) + (-\text{area B}) + (-\text{area D})$$

where area B and area D are negative in value and subtracted from area A and area C.

• If the area below the X-axis is equal to the area above the X-axis, the resulting integral is equal to zero. The graph below of f between x = a and x = b is given by:

$$\int_{a}^{b} f(x)\, dx = \text{positive region} + \text{negative region} = 0$$

where the positive region is equal to the negative region.

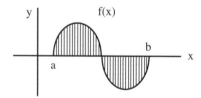

• By determining whether a *function is even or odd*, it is often possible to simplify the integral of the function to a more manageable form and solve using symmetry.

• A function is *even* if f(x) = f(−x) between x = −a and x = a. An example of a graph of an *even function* is:

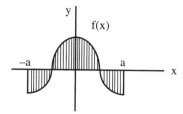

From the graph, it is clear by symmetry that the section on the left of the Y-axis between $x = -a$ and $x = 0$ is equivalent to the section on the right of the Y-axis between $x = 0$ and $x = a$. The integral for this even function can be written:

$$\int_{-a}^{a} f(x) \, dx = 2 \int_{-a}^{0} f(x) \, dx = 2 \int_{0}^{a} f(x) \, dx$$

In an even function, the area for negative values of x is equal to the area for positive values of x.

Examples of even functions include:

$$f(x) = c, f(x) = x^2, f(x) = x^4, f(x) = x^{2n}$$
$$\text{and } f((-x)^2) = (-x)(-x) = x^2.$$

• A function is *odd* if $f(x) = -f(-x)$ between $x = -a$ and $x = a$. An example of a graph of an *odd function* is:

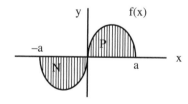

From the graph it is clear by symmetry that the section on the left of the Y-axis between $x = -a$ and $x = 0$ is equivalent but opposite to the section on the right of the Y-axis between $x = 0$ and $x = a$. The integral for this odd function can be written:

$$\int_{-a}^{a} f(x) \, dx = \int_{-a}^{0} f(x) \, dx + \int_{0}^{a} f(x) \, dx = \text{area P} + \text{area N} = 0$$

Note that $\int_{-a}^{0} f(x) \, dx = - \int_{0}^{a} f(x) \, dx$.

In an odd function, the area for negative values of x is equal but opposite to the area for positive values of x, and the two areas subtract and cancel each other out resulting in an integral equivalent to zero.

Examples of odd functions include:

$f(x) = x, f(x) = x^3, f(x) = x^5, f(x) = x^{2n+1}$
and $f((-x)^3) = (-x)(-x)(-x) = (-x)^3$

3.10 Integrating a Function and a Constant, the Sum of Functions, a Polynomial, and Properties of Integrals

• This section includes the integral of a function multiplied by a constant, the integral of the sum of functions, the integral of a polynomial function, switching limits of integration, equal limits, the integral over two subintervals, and comparing two integrals.

• The *integral of a function multiplied by a constant* is equal to the constant multiplied by the integral of the function. Therefore, if a function is multiplied by a constant or number, the constant can be brought out of the integral and multiplied with the resulting function.

$$\int_a^b C\, f(x)\, dx = C \int_a^b f(x)\, dx, \quad \text{where C is any real number.}$$

• The *graph* of the area represented by $\int_a^b f(x)\, dx$ will be elongated or narrowed along the X-axis when it is multiplied by a constant. For example:

$$\int 2\, f(x)\, dx = 2 \int f(x)\, dx$$

• The *integral of a sum of functions* is equal to the sum of the integrals. If an integral is of a sum of functions, the functions can be integrated together or as separate terms.

$$\int_a^b [f(x) + g(x)]\, dx = \int_a^b f(x)\, dx + \int_a^b g(x)\, dx$$

This can be represented using a sum:

$$\lim_{n \to \infty} \sum_{i=1}^n [f(x_i)\Delta x + g(x_i)\Delta x]$$

$$= \lim_{n \to \infty} \sum_{i=1}^n f(x_i)\Delta x + \lim_{n \to \infty} \sum_{i=1}^n g(x_i)\Delta x$$

Similarly for the *indefinite integral*:

$$\int [f(x) + g(x)]\, dx = \int f(x)\, dx + \int g(x)\, dx$$

When *constants* c_1 and c_2 are present, the sum of two functions is:

$$\int [c_1 f(x) + c_2 g(x)]\, dx = c_1 \int f(x)\, dx + c_2 \int g(x)\, dx$$

• The *integral of a polynomial function* can be evaluated term-by-term. Therefore, to integrate a polynomial function, apply the integral formula term-by-term (as with differentiating). For example:

$$\int (x^3 + x^2 + x)\, dx = \int x^3\, dx + \int x^2\, dx + \int x\, dx$$
$$= [(1/4)x^4 + c_1] + [(1/3)x^3 + c_2] + [(1/2)x^2 + c_3]$$

Combining the *constants of integration* results in:

$$(1/4)x^4 + (1/3)x^3 + (1/2)x^2 + C \quad \text{where } C = c_1 + c_2 + c_3.$$

• *Switching the limits of integration* on a definite integral reverses the sign of the integral:

$$\int_a^b f(x)\, dx = - \int_b^a f(x)\, dx$$

This occurs because integration is moving across the area in the opposite direction which reverses the sign and can be demonstrated using sums. The area is given by:

$$\int_a^b f(x)\, dx = \lim_{n\to\infty} \sum_{i=1}^{n} f(x_i)\, \Delta x = \lim_{n\to\infty} [(b-a)/n] \sum_{i=1}^{n} f(x_i)$$

where $\Delta x = (b-a)/n$.

In the negative direction, the integral is:

$$\int_{-b}^a f(x)\, dx = \lim_{n\to\infty} \sum_{i=1}^{n} f(x_i)\, (-\Delta x) = \lim_{n\to\infty} [-(a-b)/n] \sum_{i=1}^{n} f(x_i)$$

where $-\Delta x = -(a-b)/n = +(b-a)/n$.

• An integral that has both of its *limits the same* is equal to zero and represents an area over a point:

$$\int_a^a f(x)\, dx = 0$$

This integral is evaluated as:

$F(a) - F(a) = 0$

• An integral over an interval is equal to the *sum of two integrals* that represent two *subintervals* that when combined exactly equal the total original interval:

$$_a\int^c f(x)\,dx = {_a}\int^b f(x)\,dx + {_b}\int^c f(x)\,dx \quad \text{providing } a < b < c.$$

This is true because the entire area spans from $x = a$ to $x = c$, where the two areas from $x = a$ to $x = b$ and from $x = b$ to $x = c$ sum to the entire area.

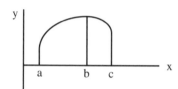

• When *comparing integrals*, the integral of function $f(x)$ is greater than or equal to the integral of function $g(x)$, if $f(x)$ is greater than or equal to $g(x)$:

$$_a\int^b f(x)\,dx \geq {_a}\int^b g(x)\,dx \quad \text{providing } f(x) \geq g(x) \text{ and } a \leq x \leq b.$$

The integrals can be compared when the area under $f(x)$ and $g(x)$ each lie on or above the X-axis and the area represented by $f(x)$ is larger or smaller than the area represented by $g(x)$.

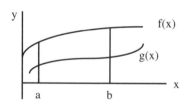

3.11 Multiple Integrals

• This section includes integrating double and triple definite and indefinite integrals; integral of a constant; and distance, velocity, and acceleration.

• Integration may be repeated multiple times. To take *multiple definite integrals* of a function, begin with the innermost integral and evaluate it at the limits of integration for the inside integral, then take the integral of the result and evaluate it at the next innermost limits of integration. Repeat this for the number of integrals specified by the number of $\int$ symbols.

• To evaluate a *double integral*: $_a\int^b \,_c\int^d f(x)\ dx\ dx$

First take the integral of $f(x)$ and evaluate it at its limits of integration c and d, then take the integral of the result and evaluate it at the limits of a and b.

• The *integral of a constant* alone is equal to the constant c_1 multiplied by the variable the integral is being integrated with respect to (indicated by dx), plus another constant c_2.

For example:

$$\int c_1\ dx = c_1 x + c_2$$

• In examples where *distance* = x(t), *velocity* = v(t), and *acceleration* = a(t), the double integral of acceleration is:

$$\iint a(t)\ dt\ dt = \int v(t) + c_1\ dt = x(t) + c_1 t + c_2$$

• To evaluate a *triple integral* of a function containing three variables:

$$_a\int^b \,_c\int^d \,_e\int^f f(x,y,z)\ dz\ dy\ dx$$

First evaluate the function f(x,y,z) by taking the integral of f(x,y,z) dz and evaluating it at the limits of e and f, next take the integral of the result with respect to dy and evaluate it at the limits of c and d, then take the integral of the result with respect to dx and evaluate it at the limits of a and b. In problems where an integral is describing volume in a coordinate system, the limits a and b, c and d, and e and f, may correspond to the X-axis, the Y-axis, and the Z-axis respectively. In this example:

$$_a\int^b \,_c\int^d \,_e\int^f f(x,\ y,\ z)\ dz\ dy\ dx$$

The first integral performed is $_e\int^f f(x,\ y,\ z)\ dz,$

the second integral is $_c\int^d$ (result of first integral) dy,

the third integral is $_a\int^b$ (result of second integral) dx.

Notice that the innermost limits on the $\int$ symbol correspond with the innermost dx, dy, or dz, and progress outward so that each $\int$ symbol corresponds with its respective dx, dy, or dz. Also note that multiple integrals may be in the forms:

$$\int \int \int f(x) \, dx \, dx \, dx$$

$$\int \int \int f(x,y,z) \, dx \, dy \, dz$$

• In an example of a triple integral where the integral of $f(x)$ is $F(x)$, the integral of $F(x)$ is $g(x)$ and the integral of $g(x)$ is $G(x)$, then the *triple integral* of $f(x)$ is:

$$\int \int \int f(x) \, dx \, dx \, dx = \int \int F(x) + c_1 \, dx \, dx$$

$$= \int g(x) + c_1 x + c_2 \, dx = G(x) + (c_1 x^2/2) + c_2 x + c_3$$

3.12 Examples of Common Integrals

• This section includes examples of selected simple *indefinite integrals* commonly used in calculus:

$$\int 0 \, dx = c$$

$$\int 2 \, dx = 2x + c$$

$$\int x^n \, dx = x^{n+1}/(n+1) + c, \quad \text{when } n \neq -1.$$

$$\int 1/x \, dx = \ln|x| + c$$

$$\int 1/(2x+3) \, dx = (1/2) \ln|2x+3| + c$$

$$\int e^x \, dx = e^x + c$$

$$\int e^{2x} \, dx = (1/2)e^{2x} + c$$

$$\int a^x \, dx = [a^x / \ln a] + c, \quad \text{when } a > 0 \text{ and } a \neq 1.$$

$$\int e^{2(x+6)} \, dx = (1/2)e^{2(x+6)} + c$$

$$\int e^u \, (du/dx) \, dx = e^{u(x)} + c$$

$\int \ln x \, dx = x \ln x - x + c,$ when $x > 0$.

$\int \cos x \, dx = \sin x + c$

$\int \sin x \, dx = -\cos x + c$

$\int \tan x \, dx = \ln|\sec x| + c$

$\int \sec x \, dx = \ln(\sec x + \tan x) + c = \log \tan(u/2 + \pi/4) + c$

$\int \csc x \, dx = \ln(\csc x - \cot x) + c = \log \tan(u/2) + c$

$\int \sec^2 x \, dx = \tan x + c$

$\int \csc^2 x \, dx = -\cot x + c$

$\int \cot x \, dx = \int (\cos x / \sin x) \, dx = |\sin x| + c$

$\int \sinh x \, dx = \cosh x + c$

$\int \cosh x \, dx = \int (\sinh x / \cosh x) \, dx = \ln(\cosh x) + c$

$\int \tanh x \, dx = \sinh x + c$

$\int [u(x) + v(x)] \, dx = \int u(x) \, dx + \int v(x) \, dx$

3.13 Integrals Describing Length

• This section develops the integral that represents the length of a curve.

• The integral can be applied to determine the *length of a curve*. Using a rectangular coordinate system, a segment of a curve or an *arc* can be determined by dividing the segment or arc into sections such that each one is nearly a straight line. From geometry, it is known that the *distance between two points* (x_1, y_1) and (x_2, y_2) is a line given by:

$$ds = \sqrt{(x_2 - x_1)^2 + (y_2 - y_1)^2}$$

Similarly, the length ds of a *section* of a curve can be represented as:

$$(\Delta s) = \sqrt{(\Delta x)^2 + (\Delta y)^2}$$

$$ds = \sqrt{(dx)^2 + (dy)^2} = \sqrt{(dx/dx)^2 (dx)^2 + (dy/dx)^2 (dx)^2}$$

$$ds = \sqrt{(dx)^2 + (dy/dx)^2 (dx)^2} = \sqrt{(dx)^2 (1 + (dy/dx)^2)}$$

$$ds = \sqrt{1 + (dy/dx)^2}\; dx$$

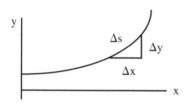

$$\Delta s = \sqrt{1 + (dy/dx)^2}\, dx$$

Therefore, the *length of an interval of a curve* can be represented as the *limit of the sum of the sections ds*. If the *length of a curve* is given by $y = f(x)$ between points $x = a$ and $x = b$, then the sum of the sections is:

$$\lim_{\Delta x \to 0} \sum \sqrt{1 + (dy/dx)^2}\, \Delta x$$

$$= \int ds = {}_a\!\int^b \sqrt{1 + (dy/dx)^2}\; dx = {}_a\!\int^b \sqrt{1 + (f'(x))^2}\; dx$$

3.14 Integrals Describing Area

• This section includes using single and double integrals to describe areas of circles, rectangular regions, regions bounded by and regions between two curves, triangular wedge regions, and surfaces of revolution.

• The integral applies to *areas of circles* as well as areas under curves. The area of a circle is πr^2 and the derivative of the area of a circle is *circumference* $2\pi r$. Conversely, the integral of the circumference of a circle $2\pi r$ is the area πr^2.

Therefore, the area of a circle $= \int (\text{circumference})\, dr$.

When applying the integral to a circle rather than a curve, the circle can be subdivided into rings of thickness Δr rather than rectangles, as depicted in the figure:

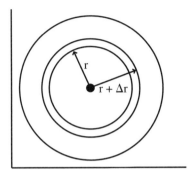

The sum of the rings can be represented by:

$$A = {_0}\!\int^r 2\pi r \, dr = \pi r^2$$

where the area of each ring is $\pi(r + \Delta r)^2 - \pi r^2$.

• *Area* can be represented using integrals of a *two-variable function* f(x,y) as well as a one-variable function f(x). Areas represented using *two variables* include *rectangles* in planes, areas in XY-planes that are *bounded by closed curves*, or areas that are two-dimensional *slices of three-dimensional objects*.

For example, a *rectangular region* can be measured within an XY-coordinate system where the *area* is defined between x = a and x = b and between y = c and y = d:

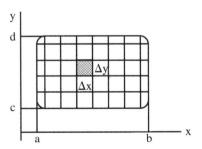

The area is divided into subintervals along each axis so that each interval has a rectangular shape and has x- and y-coordinates. The total area can be represented by the sum of this two-dimensional grid of rectangles. Each rectangle defining a subinterval of the total area has an area of $\Delta A = \Delta x \Delta y$, with $\Delta x = (b{-}a)/n$ and $\Delta y = (d{-}c)/m$, where n is the number of subdivisions along the X-axis and m is the number of subdivisions along the Y-axis. If $\Delta x_i = |x_{i+1} - x_i|$ and $\Delta y_j = |y_{j+1} - y_j|$, then the area of a subinterval can also be written $\Delta A_{ij} = \Delta x_i \Delta y_j$. The sum of all the subinterval areas as i goes from 1 to n and as j goes from 1 to m approximates the total area. As the limits of n and m approach infinity and therefore Δx and Δy approach zero, the area can be written:

$$\lim_{n\to\infty, m\to\infty} \sum_{i=1}^{n} \sum_{j=1}^{m} f(x_i, y_j)\Delta x \Delta y = \lim_{\Delta x \to 0, \Delta y \to 0} \sum_{i,j} f(x_i, y_j)\Delta x \Delta y$$

$$= \int_R f(x_i, y_j)\, dx\, dy = \int_R f\, dA$$

where R is the *region bounded by the closed curve*.

If the *area* of this region has part of the curve represented by $y = f_1(x)$ and the other part of the curve represented by $y = f_2(x)$, so that all of the curve is accounted for, the integral can be split:

$$\int_R f\, dA = \int_{f1(x)}^{f2(x)} \int_{f1(x)}^{f2(x)} dy\, dx$$

• For example, find the *area bounded by two curves*, between $y = x^2$ and $y = 1$ using a double integral:

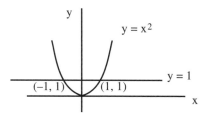

The area bounded by $y = x^2$, $y = 1$ is closed by the two intersection points $(-1,1)$ and $(1,1)$ where $x = -1$ and $x = 1$. The double integral describing these two curves is defined by where the two intersection points cross each other, which is where the two values of x and the two functions $y = f(x)$ meet:

$$\int_{-1}^{1} [1 - x^2] \int_0^1 dy\, dx = \int_{-1}^{1} [1 - x^2]\, dx = [x - x^3/3] \,\big|_{-1}^{1}$$

$$= [1 - 1/3] - [(-1) - (-1)^3/3] = [2/3] - [-1 - (-1/3)] = 4/3$$

Therefore, the area is 4/3 square units.

• If two curves represented by two functions don't intersect, *the area between the two curves* can be described by the integral of the absolute value of the difference between the two functions in a defined interval. For example, if one curve is given by f(x), the second curve is given by g(x) and the interval is a–b, the integral is: $\int_a^b |f(x) - g(x)|\, dx$

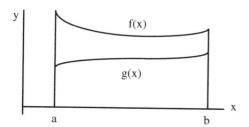

• The *area of a triangular wedge* section of a circle can be described by considering that the wedge section ($\Delta\theta/2\pi$) is a part of the whole area of the circle πr^2 and can therefore be represented as:
$(\Delta\theta/2\pi)(\pi r^2) = (1/2)r^2\Delta\theta$

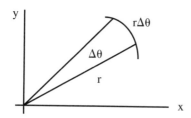

If a region contains numerous triangular wedge sections, then the total area can be represented by a sum of all the sections:

$A = \lim_{\Delta\theta \to 0} \sum (1/2)\, r^2\Delta\theta = (1/2) \int r^2\, d\theta$

• To represent *area in polar coordinates* a grid can be constructed in a similar manner to sectioning an area in rectangular coordinates and summing the subsections. (See Section 1.9 on coordinate systems.) In polar coordinates the sections are defined by r rays at various θ values where $a \leq r \leq b$ and $\alpha \leq \theta \leq \beta$, and there are n subsections in each of these directions:

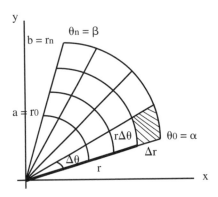

The sum of the subsections is given by: $\displaystyle\sum_{i,j} f(r_i,\theta_j)\Delta A$

where $\Delta A \approx r\,\Delta\theta\,\Delta r$ and each subsection has an area defined by Δr along the r ray multiplied by $r\Delta\theta$ along the θ direction (which is an arc).

The total area is:

$$A = \lim_{\Delta r\to 0, \Delta\theta\to 0} \sum_{i,j} f(r_i,\theta_j) r_i\Delta r\Delta\theta = \int_\alpha^\beta \int_a^b f(r,\theta)\, r\, dr\, d\theta$$

• The integral can be applied to determine *surface area* using a *surface of revolution*. If the curve of a function $y = f(x)$ is revolved around the X-axis, the surface area of the resulting surface can be determined. (Note that revolution can also occur around the Y-axis.)

• A surface that was revolved about the axis can be divided into sections similar in shape to a cylinder, where the area of an arced cylindrical surface S can be given by:

$$\Delta S = 2\pi y\,\Delta s \text{ or } dS = 2\pi y\,ds$$

The surface area of each section is the width of the cylindrical section (not the radius) multiplied by the circumference at the center of the section. Therefore, the surface area resulting from revolving curve $y = f(x)$ around the X-axis between $x = a$ and $x = b$ is:

$$S = 2\pi \int y\, ds = 2\pi \int_a^b y\, \sqrt{1+(dy/dx)^2}\, dx$$

Or approximately,

$$\sum 2\pi y \sqrt{1+(dy/dx)^2}\, \Delta x$$

where ds = $\sqrt{1+(dy/dx)^2}$ dx (described in Section 3.13.)

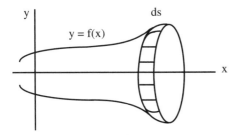

If the curve is revolved around the Y-axis instead, the surface area is:

$$S = 2\pi \int x\, ds = 2\pi \int_a^b x \sqrt{1+(dx/dy)^2}\, dy$$

Or approximately, $\sum 2\pi x \sqrt{1+(dx/dy)^2}\, \Delta y$

3.15 Integrals Describing Volume

• This section presents different techniques that may be applied to problems involving the volume of an object and includes: volume of revolution, volume by cylindrical shells, volume of spheres, volume by projecting a closed curve along the Z-axis and sectioning into columns or cubes, and volume in terms of cylindrical and spherical coordinates.

• When modeling a problem where volume must be determined, there are a variety of techniques to consider depending on the geometry of the object. The integral can also be used to define the *volume of an object*. Volume can be defined using single integral equations, double integral equations, and triple integral equations. Sums can be constructed by slicing or sectioning a three-dimensional object and adding up the sections.

• In the graph of a non-negative continuous function, the area under the curve of function y = f(x) is given by: $_a\int^b$ f(x) dx. If the function is re-volved about the X-axis between x = a and x = b, a volume is generated. This volume is called the *volume of revolution*. Each circular cross-section has an area of πy^2 and a thickness of dx. Therefore, the volume of a sec-tion is $\pi y^2 dx$. The volume of the "volume of revolution" between two vertical planes at x = a and x = b (see figure below) is given by:

$$V \approx \sum \pi y^2 \Delta x = \pi \,_a\int^b f(x)^2 \, dx = \pi \,_a\int^b y^2 \, dx$$

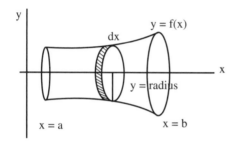

• In the graph of a non-negative continuous function between x = a and x = b, the *volume* can be described using the method of *volume by cylindrical shells*. If the area bounded by x = a and x = b is revolved about the Y-axis generating a volume, this volume can be divided along the X-axis into n parts, each having a thickness of Δx and n vertical cylinders will result. The volume of each cylindrical shell is obtained by subtracting the volume of a smaller cylinder from the next larger cylinder $\pi R^2 h - \pi r^2 h = \pi h(R^2 - r^2)$, where R is the radius of the larger cylinder and r of the smaller and h is the length along the Y-axis.

More generally, if R = (x + dx) and r = x, then subtracting the cylinders gives:

$$\pi(x + dx)^2 h - \pi x^2 h = \pi h(x + dx)(x + dx) - \pi x^2 h$$
$$= \pi h[x^2 + 2x(dx) + (dx)^2] - \pi x^2 h$$
$$= \pi h x^2 + 2x\pi h(dx) + \pi h(dx)^2 - \pi x^2 h = 2x\pi h(dx) + \pi h(dx)^2$$

If the sum of the n shells is taken as n approaches infinity and the thick-ness of each shell approaches zero, the $\pi h(dx)^2$ term will quickly approach zero because of $(dx)^2$ or Δx^2. The volume depicted below can be described by: $V \approx \sum 2\pi x h \, \Delta x$

As the number of shells approaches infinity, then the volume becomes:

$$V = {}_a\!\int^b 2\pi hx \; dx$$

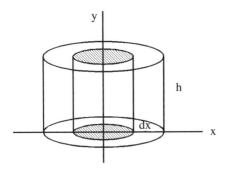

• The integral can also be applied to the *volume of spheres*. The volume of a sphere can be divided up into nested spheres or shells, each having a thickness of Δr, where each shell can be measured from radius r to radius $r + \Delta r$. The measurement of each shell can be estimated by its surface area multiplied by Δr, or $4\pi r^2\Delta r$. Therefore, the volume is the sum of the volumes of the incremental shells:

$$V = {}_0\!\int^r 4\pi r^2 \; dr = (4/3)\pi r^3$$

which is the volume of a sphere.

• *Volume* of an object can be described by projecting a closed curve vertically along the Z-axis in an XYZ coordinate system into a three-dimensional solid. See figures below. This volume can be determined in terms of double or triple integration and summation. If the volume of this object is divided into *columns* in the direction of the Z-axis, where an area subdivision dydx in the XY plane is projected vertically along the Z-axis to the surface $z = F(x,y)$, the volume of each column is $F(x,y)$ dydx. The sum of all the columns as the number of columns approaches infinity, gives the total volume for the function $F(x,y)$ and is described by:

$$V = \lim_{\Delta x \to 0, \Delta y \to 0} \sum_{i,j} F(x_i, y_j)\Delta x\Delta y = \iint_R F(x_i, y_j) \; dxdy = \iint_R F \; dA$$

where the " $_R$ " defines a region of area.

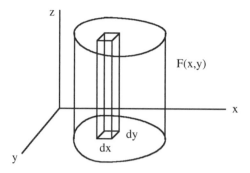

An alternative method of describing this same volume is to divide it into *cubes* in the XYZ coordinate system so that the volume of each cube is dxdydz. Then the sum of all the cubes gives the total volume and can be described by:

$$V = \int_a^b \int_{f1(x)}^{f2(x)} \int_{F1(x,y)}^{F2(x,y)} dz\, dy\, dx$$

where $F_1(x,y) = z$ and $F_2(x,y) = z$ describe the surfaces along z and $f_1(x) = y$ and $f_2(x) = y$ describe the closed curve. The interval along the X-axis is from x = a to x = b.

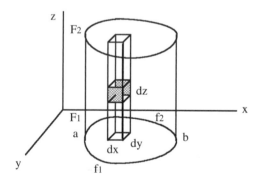

• In general, a volume in an XYZ coordinate system can be divided into cubes, such that each cube has a volume $\Delta x \Delta y \Delta z = \Delta V$ and $\Delta x = (b - a)/m$, $\Delta y = (c - d)/n$, $\Delta z = (e - f)/p$ and m, n, and p correspond to the number of subdivisions along the three axes. The sum of all the cubes gives the total volume and is described along the three axes by:

$$V = \int_a^b \int_c^d \int_e^f dz\, dy\, dx$$

• The *volume of a rectangular* solid can be determined by dividing it into many cubic or rectangular sections. Consider a rectangular solid positioned with its lower left corner at the vertex of an XYZ coordinate system:

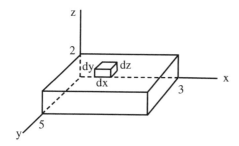

The rectangular solid spans from x = 0 to x = 3, y = 0 to y = 5, and z = 0 to z = 2. The integral describing volume is:

$$V = {_0\!\int^3} {_0\!\int^5} {_0\!\int^2} dz\ dy\ dx = {_0\!\int^3} {_0\!\int^5} 2\ dy\ dx = {_0\!\int^3} 2y\big|_0^5\ dx$$

$$= {_0\!\int^3} 10\ dx = 10x\big|_0^3 = 30 \text{ cubic units}$$

In this example, the volume can also be obtained by simple geometry as length × width × height, or

3 × 5 × 2 = 30 cubic units.

• In *cylindrical coordinates*, the volume of an object can be sectioned in a grid of subsections where the volume of each subsection is defined according to the coordinates. Then the subsections can be summed or integrated to find the total volume. (See Section 1.9 on coordinate systems.)

In cylindrical coordinates the sections are defined by r rays at various θ values and projected along the Z-axis with a $\leq$ r $\leq$ b, $\alpha \leq \theta \leq \beta$ and c $\leq$ z $\leq$ d. The r component is measured from the Z-axis, the θ component measures the distance around the Z-axis and the z component measures along the Z-axis.

The following figure depicts a point $P(r,\theta,z)$ in a cylindrical coordinate system:

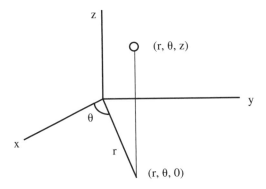

Where $a \leq r \leq b, \alpha \leq \theta \leq \beta, c \leq z \leq d$

- For example, the *volume of a triangular wedge* section that is projected vertically along the Z-axis can be represented in *cylindrical coordinates*.

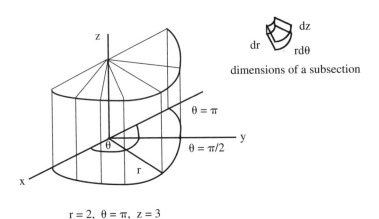

dimensions of a subsection

$r = 2, \; \theta = \pi, \; z = 3$

The area of each subsection is $\Delta A \approx r \, \Delta\theta \, \Delta r$ and is defined by Δr along the r-ray multiplied by $r \, \Delta\theta$ along the θ direction. Therefore, the volume of a subsection is $\Delta V \approx r \, \Delta r \, \Delta\theta \, \Delta z$. For the total volume containing numerous triangular wedge sections where $r = 2, \theta = \pi$, and $z = 3$, the volume can be represented by:

$$V = \int_c^d \int_\alpha^\beta \int_a^b r \, dr \, d\theta \, dz = \int_0^3 \int_0^\pi \int_0^2 r \, dr \, d\theta \, dz$$

$$= \int_0^3 \int_0^\pi 2^2/2 \, d\theta \, dz = \int_0^3 \int_0^\pi 2 \, d\theta \, dz = \int_0^3 2\pi \, dz = 6\pi$$

To verify this answer, consider that both the top and bottom are level, therefore the volume should be equivalent to the area of one end multiplied by the z-dimension:

$$(\Delta\theta/2\pi)\pi r^2(z) = (1/2)r^2\Delta\theta(z) = (1/2)(4)\pi(3) = 6\pi$$

where $(\Delta\theta/2\pi)\pi r^2$ defines a wedge of a circle.

- To represent *volume in spherical coordinates*, where $x = \rho \cos \theta \sin \phi$, $y = \rho \sin \theta \sin \phi$, $z = \rho \cos \phi$, $\rho = (x^2 + y^2 + z^2)^{1/2}$, a grid can be constructed in a similar manner to sectioning a volume in rectangular coordinates and summing the subsections. (See Section 1.9 on coordinate systems.) In spherical coordinates, the sections are expressed in terms of ρ, θ, and ϕ, where ρ can range from 0 to ∞ and originates from the origin, θ can range from 0 to 2π and measures the distance around the Z-axis, and ϕ can range from 0 to π and measures down from the Z-axis. Note that ρ is measured from the origin rather than the Z-axis as is the case with r in cylindrical coordinates. This figure depicts a point $P(\rho,\theta,\phi)$ in a spherical coordinate system:

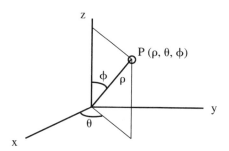

To find volume in spherical coordinates, divide the object into n subsections in each of the ρ, θ, and ϕ directions and integrate the volume of each subsection over $0 \le \rho \le \infty$, $0 \le \theta \le 2\pi$, and $0 \le \phi \le \pi$. Each subsection is a semi-rectangular volume element and is defined in terms of its (ρ,θ,ϕ) coordinates as depicted on the next page. One edge of the element has a length of $\Delta\rho$, one edge is defined by rotating ρ the length $\Delta\phi$ resulting in length $\rho \Delta\phi$, and one edge is defined by rotating in the θ direction measured out from the Z-axis at the length $(\rho \sin \phi)$ resulting in a length of $(\rho\sin \phi \Delta\theta)$. Therefore, the volume of a subsection is:

$$\Delta V = (\Delta\rho)(\rho\Delta\phi)(\rho \sin \phi \Delta\theta) = \rho^2(\sin \phi) \Delta\rho \Delta\theta \Delta\phi$$

The total volume is the sum of the subsections as the size of each subsection approaches zero, which is the integral:

$$V = {}_0\!\int^\pi {}_0\!\int^{2\pi} {}_0\!\int^\infty \rho^2 \sin\phi \, d\rho \, d\theta \, d\phi$$

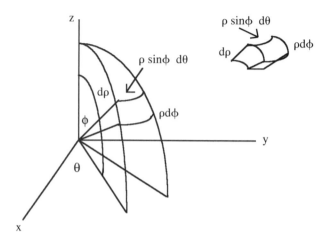

• For example, the volume of a *sphere* located at the origin of a spherical coordinate system, having a radius = R is given by:

$$V = {}_0\!\int^\pi {}_0\!\int^{2\pi} {}_0\!\int^R \rho^2 \sin\phi \, d\rho \, d\theta \, d\phi$$

$$= {}_0\!\int^\pi {}_0\!\int^{2\pi} (1/3)R^3 \sin\phi \, d\theta \, d\phi = {}_0\!\int^\pi (2\pi)(1/3)R^3 \sin\phi \, d\phi$$

$$= (2\pi)(1/3)R^3 \, [-\cos\pi - -\cos 0] = (2/3)\pi R^3[+1 + 1]$$

$$= (4/3)\pi R^3$$

which is the known volume of a sphere.

3.16 Changing Coordinates and Variables

• This section provides a brief summary of changing coordinates and variables including the volume of a sphere in rectangular, cylindrical, and spherical coordinates, and changing between coordinates using the Jacobian factor.

• When describing areas and volumes of circles, spheres, cylinders, and other non-rectangular shapes, it may be advantageous to evaluate integrals in polar or spherical coordinates rather than rectangular coordinates.

- For example, the *volume of a sphere* located at the origin given in Cartesian (rectangular), cylindrical, and spherical coordinates can be represented as follows:

(a.) In *rectangular coordinates*, the *equation for a sphere* is, $x^2 + y^2 + z^2 = R^2$. The sphere can be divided into sections such as octants with the volume of each octant further divided into small rectangular or cubic subsections. In an XYZ coordinate system, the volume of each cube is dxdydz. Then the sum of all the cubes in the octant multiplied by 8 gives the total volume of the sphere. Using this strategy, the volume of a sphere located at the origin of the coordinate system can be represented by:

$$V = (8) \int_0^b \int_0^{\sqrt{R^2-x^2}} \int_0^{\sqrt{R^2-x^2-y^2}} dz \, dy \, dx = (4/3)\pi R^3$$

The interval along the X-axis is from $x = 0$ to $x = b$.

(b.) In *cylindrical coordinates*, the *equation for a sphere* is, $r^2 + z^2 = R^2$. Again, the sphere can be divided into sections such as octants with the volume of each octant further divided into small subsections. If the sphere is located at the origin, the equation for volume is:

$$V = (8) \int_0^{\sqrt{R^2-r^2}} \int_0^{\pi/2} \int_0^b r \, dr \, d\theta \, dz = (4/3)\pi R^3$$

(c.) In *spherical coordinates*, the *equation for a sphere* is, $\rho = R$. If the sphere is divided into octants with the volume of each octant further divided into small subsections, and the sphere is located at the origin, then volume can be represented as:

$$V = (8) \int_0^{\pi/2} \int_0^{\pi/2} \int_0^R \rho^2 \sin \phi \, d\rho \, d\theta \, d\phi = (4/3)\pi R^3$$

- To *change between coordinate systems* when evaluating an integral, the expressions for x, y, and z can be substituted from the original coordinates to the new coordinates. For example, to make a simple change from rectangular to polar coordinates for area, first substitute $x = r \cos \theta$ and $y = r \sin \theta$ in the *integrand*, then change the *limits of integration* to describe the dimensions within the new coordinate system. Finally, replace dA with r drdθ.

- A general method used to *change variables* uses the *Jacobian J* determinant for two or three variables. In this method, x, y, and z are related to new coordinates u, v, and w, such that $x = x(u,v,w)$, $y = y(u,v,w)$, and $z = z(u,v,w)$. The Jacobian represents a *factor* that relates the original coordinate system to the new coordinate system.

For a *single integral*, changing from:

$$\int f(x)\ dx \text{ to } \int f(u)\ du,$$

the *factor* relating dx and du is simply the ratio dx/du. Therefore, the dx can be replaced with (dx/du)du. Similarly, (du/dx)dx is equivalent to du.

For a *double integral*, changing from:

$$\iint f(x,y)\ dx\ dy \text{ to } \iint f(u,v)\ du\ dv,$$

the *factor* given by J relates the area dxdy with area dudv such that dxdy becomes $|J|$ dudv. The variables x and y are related in that x = x(u,v) and y = y(u,v), and each point in the x–y coordinate system is related to each point in the u–v coordinate system. Therefore, the integral in the x–y system corresponds to the integral in the u–v system as:

$$\iint f(x,y)\ dx\ dy = \iint f(x(u,v),y(u,v))|J|\ du\ dv.$$

In a *triple integral*, when changing from:

$$\iiint f(x,y,z)\ dx\ dy\ dz \text{ to } \iiint f(u,v,w)\ du\ dv\ dw,$$

the factor relates the volume dxdydz with volume dudvdw such that dxdydz becomes $|J|$ dudvdz. Therefore, the integral in the x–y–z system corresponds to the integral in the u–v–w system as:

$$\iiint f(x,y,z)\ dxdydz = \iiint f(x(u,v,w),y(u,v,w),z(u,v,w))|J|\ dudvdw$$

• The *J factor* in two and three variable integrals, represents the *Jacobian determinant* and corresponds to the (dx/dy) factor in the one-variable integral. There are both two- and three-dimensional Jacobian determinants that correspond to two- and three-variable integrals. The *Jacobian determinant* for *two variables* is:

$$J = \begin{vmatrix} \partial x/\partial u & \partial x/\partial v \\ \partial y/\partial u & \partial y/\partial v \end{vmatrix} = \frac{\partial x}{\partial u}\frac{\partial y}{\partial v} - \frac{\partial y}{\partial u}\frac{\partial x}{\partial v}$$

The *Jacobian determinant* for *three variables* is:

$$J = \begin{vmatrix} \partial x/\partial u & \partial x/\partial v & \partial x/\partial w \\ \partial y/\partial u & \partial y/\partial v & \partial y/\partial w \\ \partial z/\partial u & \partial z/\partial v & \partial z/\partial w \end{vmatrix} = \frac{\partial x}{\partial u}[\frac{\partial y}{\partial v}\frac{\partial z}{\partial w} - \frac{\partial z}{\partial v}\frac{\partial y}{\partial w}]$$

$$-\frac{\partial y}{\partial u}[\frac{\partial x}{\partial v}\frac{\partial z}{\partial w} - \frac{\partial z}{\partial v}\frac{\partial x}{\partial w}]+\frac{\partial z}{\partial u}[\frac{\partial x}{\partial v}\frac{\partial y}{\partial w} - \frac{\partial y}{\partial v}\frac{\partial x}{\partial w}]$$

These determinants for two and three variables are sometimes represented by:

$$\frac{\partial(x,y)}{\partial(u,v)} \quad \text{and} \quad \frac{\partial(x,y,z)}{\partial(u,v,w)}$$

Also note that the ∂ represents a *partial derivative*. See Chapter 6 for a discussion of partial derivatives.

• The *J factor* represented by the Jacobian determinant can be derived by translating an element of the area or volume represented by a two- or three-variable integral from an XY (or XYZ) coordinate system to a UV (or UVW) coordinate system. Consider the relationship of a volume element in an XYZ coordinate system vs. a UVW coordinate system that represents a cylindrical or spherical system. Because a curved region in a rectangular coordinate system can correspond to a rectangular region in cylindrical or spherical coordinates, the volume element has curved sides in the rectangular coordinates and straight sides in cylindrical and spherical coordinates.

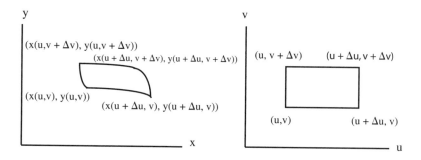

This figure represents only the x–y and u–v planes for simplicity. If the curved element in the XYZ coordinate system is represented in vector form such that the vector representing each curved line has an **i** component in the x-direction, a **j** component in the y-direction, and a **k** component in the z-direction (see Section 5.1 for explanation of **i**, **j**, **k**,), then:

The change in x is given by:

$$(x(u+\Delta u, v, w) - x(u, v, w))\mathbf{i} + (y(u+\Delta u, v, w) - y(u, v, w))\mathbf{j}$$
$$+ (z(u+\Delta u, v, w) - z(u, v, w))\mathbf{k}$$

$$\approx [(\partial x/\partial u)\Delta u]\mathbf{i} + [(\partial y/\partial u)\Delta u]\mathbf{j} + [(\partial z/\partial u)\Delta u]\mathbf{k}$$

The change in y is given by:

$$(x(u, v+\Delta v, w) - x(u, v, w))\mathbf{i} + (y(u, v+\Delta v, w) - y(u, v, w))\mathbf{j}$$
$$+ (z(u, v+\Delta v, w) - z(u, v, w))\mathbf{k}$$

$$\approx [(\partial x/\partial v)\Delta v]\mathbf{i} + [(\partial y/\partial v)\Delta v]\mathbf{j} + [(\partial z/\partial v)\Delta v]\mathbf{k}$$

The change in z is given by:

$$(x(u, v, w+\Delta w) - x(u, v, w))\mathbf{i} + (y(u, v, w+\Delta w) - y(u, v, w))\mathbf{j}$$
$$+ (z(u, v, w+\Delta w) - z(u, v, w))\mathbf{k}$$

$$\approx [(\partial x/\partial w)\Delta w]\mathbf{i} + [(\partial y/\partial w)\Delta w]\mathbf{j} + [(\partial z/\partial w)\Delta w]\mathbf{k}$$

The vector product (cross product) of the x, y, and z components is:

$$\Delta u \Delta v \Delta w \times$$

$$\left| \frac{\partial x}{\partial u} \left[\frac{\partial y}{\partial v} \frac{\partial z}{\partial w} - \frac{\partial z}{\partial v} \frac{\partial y}{\partial w} \right] - \frac{\partial y}{\partial u} \left[\frac{\partial x}{\partial v} \frac{\partial z}{\partial w} - \frac{\partial z}{\partial v} \frac{\partial x}{\partial w} \right] + \frac{\partial z}{\partial u} \left[\frac{\partial x}{\partial v} \frac{\partial y}{\partial w} - \frac{\partial y}{\partial v} \frac{\partial x}{\partial w} \right] \right|$$

where the quantity within the absolute value symbol represents the determinant:

$$J = \begin{vmatrix} \partial x/\partial u & \partial x/\partial v & \partial x/\partial w \\ \partial y/\partial u & \partial y/\partial v & \partial y/\partial w \\ \partial z/\partial u & \partial z/\partial v & \partial z/\partial w \end{vmatrix} = \frac{\partial(x,y,z)}{\partial(u,v,w)}$$

• In general, to change coordinate systems in an integral, express x, y, and z in terms of the new variables u, v, and w, convert the x–y–z element into a u–v–w element, and include the Jacobian determinant to factor in the change in the shape of the element.

• The Jacobian determinants can derive the conversion factors for polar, cylindrical, and spherical coordinate systems.

In *polar coordinates*:

$$x = r \cos \theta = u \cos v$$
$$y = r \sin \theta = u \sin v$$

The *Jacobian in polar coordinates* is:

$$J = \begin{vmatrix} \partial x/\partial r & \partial x/\partial \theta \\ \partial y/\partial r & \partial y/\partial \theta \end{vmatrix} = \begin{vmatrix} \cos\theta & -r\sin\theta \\ \sin\theta & r\cos\theta \end{vmatrix}$$

$$= r \cos^2\theta + r \sin^2\theta = r$$

In *spherical coordinates*:

$$x = \rho \cos \theta \sin \phi = u \cos v \sin w$$
$$y = \rho \sin \theta \sin \phi = u \sin v \sin w$$
$$z = \rho \cos \phi = u \cos w$$

The *Jacobian in spherical coordinates* is:

$$J = \begin{vmatrix} \partial x/\partial \rho & \partial x/\partial \phi & \partial x/\partial \theta \\ \partial y/\partial \rho & \partial y/\partial \phi & \partial y/\partial \theta \\ \partial z/\partial \rho & \partial z/\partial \phi & \partial z/\partial \theta \end{vmatrix}$$

$$= \begin{vmatrix} \cos\theta \sin\phi & \rho\cos\theta\cos\phi & -\rho\sin\theta\sin\phi \\ \sin\theta\sin\phi & \rho\sin\theta\cos\phi & \rho\cos\theta\sin\phi \\ \cos\phi & -\rho\sin\phi & 0 \end{vmatrix}$$

Calculating this determinant results in $J = \rho^2 \sin \phi$.

3.17 Applications of the Integral

• This section includes some applications of the integral including work, pressure, center of mass, and distributions.

Work and Pressure

• *Work* performed = force × distance. For example, if a particle or object is moved by a constant force F some distance x, then the *work* done is W = (F)(x). For a *variable force* F pushing or pulling a particle or object in the direction of motion along a straight line from x_1 to x_2, work can be represented by:

$$_{x1}\!\int^{x2} F(x)\, dx$$

The motion is often described along an axis of a coordinate system. If the *motion is along a curve* ds from s_1 to s_2, then the total work done can be represented by:

$$_{s1}\!\int^{s2} F(s)\, ds$$

• **Example:** What is the work required to lift an object weighing 1,000 pounds six feet off the ground?

$$\int_0^6 (1,000)\, dx = (1,000)(6) - (1,000)(0) = 6,000 \text{ foot-pounds}$$

(Note that pounds is the weight. If mass were given it would need to be multiplied with the acceleration of gravity in the proper units.)

Because this is a simple problem that does not require sectioning into subsections or elements and then summing as the size of each subsection approaches zero, this result of integration can be compared with simply multiplying.

W = (force)(distance)
W = (1,000 pounds)(6 feet) = 6,000 foot-pounds

• In more complicated work integrals, the geometry of what is being described is written into the integral. For example, to calculate the work required to pump fluid out of a cylindrical tank of height h, the integral could be designed to calculate the sum of disc-shaped sections of water that each need to be lifted out of the tank. The volume of each disc is $\pi r^2 dz$ and the weight of a disc is the density ρ of the fluid multiplied by the volume of the disc, or $\rho \pi r^2 dz$. (This is not the same ρ as described in polar coordinate systems.) If the distance each disc needs to be lifted is given by z (which will be slightly different for each disc because they are starting from different heights), the integral giving weight multiplied with distance is:

$$\int_0^h \rho \pi r^2 z\, dz$$

• *Pressure* is force per unit area, P = F/A, and therefore force is pressure multiplied with area, F = PA. The pressure exerted by a liquid at a given point in the liquid or on an object that is submersed in the liquid, is the same in all directions at a given point. The pressure increases the deeper it is measured in a body of liquid and is given by:

P = ρgh = wh

where ρ is the density of the liquid (mass/volume);

g is the gravitational constant;

h is the height, or more specifically depth, from the surface of the liquid; and

w is the weight per volume of the liquid, or ρg.

The total pressure on an object or on a specified region of liquid can be represented as the sum of all the subdivisions of area at various depths. If a section of area is defined by the length of a horizontal section l at a given height h with a thickness dh, then the force on the section is:

$$\Delta F = whl\, \Delta h$$

Therefore, the total *force* F on the object or region is the sum of the forces of all sections:

$$F = \int whl\, dh$$

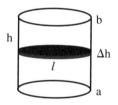

Consider the *work* required to pump water up and out of a tank that has a height of h and a diamter of l. If the tank is divided into sections having area A, thickness Δh, and each at a slightly different height, then this can be thought of as pumping each section out individually and summing:

$$W = {}_a\!\int^b Awh\, dh$$

Center of Mass

• Integrals can be used to describe the *center of mass*. The center of *mass* is defined as the *moment* divided by the mass, where the moment is defined as the distance of the mass from a line or axis multiplied by the mass itself. For example, for a point mass m_1 located a certain distance x_1 from an axis, the moment of that point mass is $x_1 m_1$. If there are many point masses located at specific distances from the axis, the sum of those moments is: $\sum x_i m_i$

If there is a continuous distribution of these masses at various lengths from the axis, the sum of the moments becomes: $\int x\, dm$

The center of mass is: $\int x\, dm \div \int dm$

where dm represents an element of mass.

• The *center of mass* can be measured from an axis or a plane in terms of one, two, or three dimensions. In one dimension, masses can behave as a single mass along a line or curve, such that the density ρ of each element of mass is described in terms of ρ = mass/volume, and the volume is represented in this one-dimensional situation as the length of the element of mass, so that the density at a point is dm/ds. Therefore, the mass is density multiplied by length, $\rho\, ds$. The total center of mass is:

$$\text{moment/mass} = \int \rho x\, ds \div \int ds$$

In an XY coordinate system, the *center of mass* described according to x-and y-coordinates is:

$$\int \rho x\, ds \div \int ds \text{ and } \int \rho y\, ds \div \int ds$$

• For two-dimensions, *center of mass* is described in terms of density multiplied by area such that $dm = \rho\, dA$, and the masses are distributed in a plane rather than along a line or curve. The center of mass occurs where all the masses balance. There is a moment around the X-axis, $\Sigma y_i m_i$, and the Y-axis, $\Sigma x_i m_i$. Each element has x and y coordinates within the plane. Therefore:

$$dm = \rho\, dA = \rho(y_1 - y_2)\, dx = \rho\, (x_1 - x_2)\, dy$$

In an X-Y coordinate system, the center of mass described according to x- and y-coordinates is:

$$\int \rho(y_1 - y_2)\, x\, dx \div \int \rho(y_1 - y_2)\, dx \text{ and}$$
$$\int \rho(x_1 - x_2)\, y\, dy \div \int \rho(x_1 - x_2)\, dy$$

• For three-dimensions, *center of mass* is described in terms of density multiplied by volume such that $dm = \rho\, dV$, and the masses are distributed in three-dimensional space. If the volume is described in terms of a volume of revolution:

$$V = \pi \int_a^b f(x)^2\, dx \text{ or } V = \pi \int_a^b y^2\, dx$$

and sections are made perpendicular to the Y-axis, then:

$$dm = \rho\pi x^2\, dy$$

Each section is the same distance from the bottom plane, so that a first moment is given by:

$$\int \rho\pi x^2\, y\, dy$$

The center of mass is described as the moment divided by the mass:

$$\int \rho x^2 y \, dy \div \int \rho x^2 \, dy$$

• Note that *moment of inertia* is used in the description of physical systems such as rotating bodies. The moment of inertia is calculated similar to
the first moment in the center of mass description above, except that the distance from the mass to the axis or plane is squared. For example, in the one-dimensional situation the first moment is:

$$\int \rho x \, ds \text{ and } \int \rho y \, ds$$

Using the square of distance, the *moment of inertia* is:

$$\int \rho x^2 \, ds \text{ and } \int \rho y^2 \, ds$$

Distributions, Probabilities, and Integration

• Integrals are often used to describe various statistical quantities. One example is the *probability* that some value of x will fall between two points, a and b, and is described in terms of a *density function* p(x) and the integral:

$$_a\!\int^b p(x) \, dx$$

Providing $_{-\infty}\!\int^{\infty} p(x) \, dx = 1$ and $p(x) \geq 0$ for all x values.

This can be represented as an *area* between points a and b.

• An example using two variables occurs where the two variables are distributed throughout a population. The density function can be given in terms of two variables p(x,y) and a volume is described on a graph such that x is between x = a and x = b and y is between y = c and y = d:

$$_a\!\int^b {}_c\!\int^d p(x,y) \, dy dx$$

Providing $_{-\infty}\!\int^{\infty} {}_{-\infty}\!\int^{\infty} p(x,y) \, dy \, dx = 1$ and $p(x) \geq 0$ for all x and y.

3.18 Evaluating Integrals Using Integration by Parts

• This section provides a summary of the method of integration by parts, which is a practical method used often to evaluate integrals. Included in this section is the formula for integration by parts for indefinite and definite integrals and its derivation.

• To evaluate complicated integrals, methods that go beyond simply applying the integral formula are often required. Some of the most common methods are integration by parts, substitution, partial fractions, and looking up integrals in integral tables.

• To evaluate certain complicated integrals, the method of *integration by parts* can be applied. Applying this method is often compared to applying the *product rule* for evaluating derivatives. To use this method, the integral must exist in the form or be arranged to fit the following formula:

$$\int f(x)\, g'(x)\, dx = f(x)\, g(x) - \int f'(x)\, g(x)\, dx$$

Or equivalently:

$$\int f\, g'\, dx = f\, g - \int f'\, g\, dx$$

Note: The integration by parts formula is an important formula used frequently in calculus.

Using other notation, where u = f(x) and v = g(x), the integration by parts formula is written:

$$\int u\, dv = uv - \int v\, du$$

• The integration by parts formula can be derived by integrating the product rule. The *product rule* is:

$$\frac{d}{dx} f(x) g(x) = g(x) \frac{d}{dx} f(x) + f(x) \frac{d}{dx} g(x)$$

or (fg)' = f'g + fg'

Integrate each term:

$$\int (fg)'\, dx = \int f'g\, dx + \int fg'\, dx$$

Rearrange:

$$\int fg'\, dx = \int (fg)'\, dx - \int f'g\, dx$$

Because $\int (fg)'\, dx = fg$, this becomes:

$$\int fg'\, dx = fg - \int f'g\, dx$$

which is the integration by parts formula.

• When the integral is a *definite integral*, the f(x)g(x) term is also evaluated at the limits, and the integration by parts formula for a definite integral becomes:

$$_a\!\int^b f(x)\, g'(x)\, dx = f(b)g(b) - f(a)g(a) - {}_a\!\int^b f'(x)\, g(x)\, dx$$

• To use integration by parts, the appropriate parts of the integral must be substituted for f'(x) and g(x), (or equivalently u and dv). The choice should be dependent on how easy it will be to get f(x) from f'(x).

• **Example:** Integrate $\int x \cos x\, dx$.

First arrange in a form that will fit the integration by parts formula and make the following substitutions:

u = x
du = dx
dv = cos x dx

From the substitution for dv, integrate v:

$$v = \int \cos x\, dx = \sin x$$

Substituting into the integration by parts formula:

$$\int u\, dv = uv - \int v\, du$$

The integral becomes:

$$\int x \cos x\, dx = x \sin x - \int \sin x\, dx$$

Because $\int \sin x\, dx = -\cos x$, the integral becomes:

$$\int x \cos x\, dx = x \sin x + \cos x + c$$

• The integration by parts formula may need to be *repeated* during the evaluation of an integral.

For example, integrate: $\int x^2 e^x\, dx$:

Using the integration by parts formula: $\int u\, dv = uv - \int v\, du$

The substitutions are:

$$u = x^2, dv = e^x \, dx, \text{ therefore, } v = e^x \text{ and } du = 2x \, dx$$

Substitute into the formula:

$$\int x^2 e^x \, dx = x^2 e^x - \int e^x \, 2x \, dx$$

Repeat integration by parts on $\int e^x \, 2x \, dx$

The substitutions are:

$$u = 2x, dv = e^x \, dx, \text{ therefore, } v = e^x \text{ and } du = 2 \, dx$$

Substitute into the formula:

$$\int e^x \, 2x \, dx = 2xe^x - 2\int e^x \, dx = 2xe^x - 2e^x + c$$

Therefore, substituting back into the original integral:

$$\int x^2 e^x \, dx = x^2 e^x - (2xe^x - 2e^x) + c = e^x(x^2 - 2x + 2) + c$$

(See the end of Section 3.19 for simplifications that can be used to simplify integrals before applying integration by parts. These include using trigonometric identities, factoring, and some common derivatives.)

3.19 Evaluating Integrals Using Substitution

• This section provides a summary of the method of substitution, which is a practical method used often to evaluate integrals. Included in this section is a description of substitution, its relationship to the chain rule, using substitution for indefinite and definite integrals, and simplifying integrals before applying substitution using trigonometric identities, factoring, and common derivatives.

• *Substitution of variables* is used to translate a complicated integral into a more manageable form so that the integral can be solved using the integral formula or integral tables. Then the integral is translated back to its original variables. When using substitution or other methods, there are certain types of simplifications that may be particularly helpful to perform on the original integral before the more formal technique is employed. Such simplifications may involve using trigonometric identities, the Pythagorean theorem, and factoring.

• The method of substitution is sometimes referred to as *change of variables*. The substitution method is a general strategy involving substituting placeholder variables into a complicated integral, solving the integral in its simplified form, then substituting the original variables back into the resulting expression.

• Using substitution to solve integrals is often compared with using the *chain rule* to evaluate derivatives. In fact the substitution method is generally based on the chain rule from differentiation. To understand substitution it is helpful to review the chain rule. The chain rule is used often to determine the derivative of *composite functions*. To use the chain rule for differentiation, it is important to identify the outer function and the inner function in the equation or expression to be differentiated. The chain rule formula is:

$$[f(g(x))]' = f'(g(x))(g'(x))$$

Or equivalently, for more complex functions:

$$\frac{d}{dx}(f(x))^n = n \times (f(x))^{n-1} \times \frac{d}{dx}f(x)$$

The chain rule essentially states that the derivative of $f(g(x))$ equals the derivative of the outside function multiplied by the derivative of the inside function. For example, use the chain rule to differentiate:

$$(d/dx)(x^2 + 7)^3 = 3(x^2 + 7)^2(2x)$$

The chain rule results in the product of two factors, the derivative of the outside function and the derivative of the inside function. If the integral in question has a form similar to $f'(g(x))(g'(x))$, then its integral or antiderivative has the form $f(g(x))$.

• Substitution can generally be applied to integrals that have their integrand in a form similar to $f'(g(x))(g'(x))$. The key to using the substitution method is identifying the "inside function" by looking at which factor is the derivative of the inside function. This generally involves guessing what the antiderivative could be by using the reverse of the chain rule and trying to identify the inside function and the outside function from the factors, then differentiating that result to check whether it was, in fact, the antiderivative. The substitution method may produce a result that is off by a constant factor which can be corrected, then checked by differentiation.

• A general form of an integral that lends itself to substitution is $f(g(x))(g'(x)) = f(u)(g'(x))$. If $f(u)$ is a continuous function and $g(x)$ is a function such that $dg(x)/dx$ exists and $g(x) = u = u(x) =$ the inside function, then:

$$\int f(u)\, du = \int f(g(x))g'(x)\, dx$$

Equivalently:

$$\int f(u)\, du = \int f(u(x))\frac{du(x)}{dx}\, dx$$

where $du = (du/dx)dx$

• To use the substitution method to calculate an integral, first identify the inside function by looking at which factor is the derivative of the inside function, choose u and compute du/dx, identify and integrate f(u) du, then substitute u back into the antiderivative.

• **Example:** Find the integral $\int f(x)\, dx = \int \left(\dfrac{x+1}{x^2+2x+10}\right)dx.$

First, identify the inside function by looking at which factor is the derivative of the inside function.

Notice that the numerator (x + 1) is the derivative of the denominator (x² + 2x + 10), except for a factor of 2. Therefore, choose u = (x² + 2x + 10).

The derivative of u is: du/dx = (2x + 2) = 2(x + 1)

Rearranging gives: dx = du / 2(x + 1)

Next, substitute u and du into the integral so that (x² + 2x + 10) = u and dx = du / 2(x + 1), then evaluate the integral:

$$\int \frac{(x+1)du}{(u)(2)(x+1)} = \int \frac{1}{2u}\, du = \frac{1}{2}\int \frac{1}{u}\, du = \frac{1}{2}\ln|u| + c$$

(Remember, $\int 1/x\, dx = \ln|x| + c$.)

Finally, substitute the original expressions back into the evaluated integral, where u = (x² + 2x + 10):

(1/2) ln$|$(x² + 2x + 10)$|$ + c

Therefore:

$\int$(x + 1)/(x² + 2x + 10) dx = (1/2) ln$|$(x² + 2x + 10)$|$ + c

- **Example:** Find the integral $\int$ f(x) dx = $\int$ x(exp{x²}) dx.

First, identify the inside function by looking at which factor is the derivative of the inside function.

Notice that the factor x is the derivative of x², except for a factor of 2. Therefore choose u = x².

The derivative of u is: du/dx = 2x, or du = 2x dx

Rearranging gives: dx = du/2x

Next, substitute u and du into the integral so that x² = u and dx = du/2x, then evaluate the integral:

$\int$ x (exp{u}) du/2x = (1/2)$\int$ exp{u} du = (1/2) exp{u} + c

Finally, substitute the original expressions back into the evaluated integral, where u = x²:

(1/2) exp{x²} + c

Therefore, $\int$ x (exp{x²}) dx = (1/2) exp{x²} + c.

- If the integral is a *definite integral*, the limits can be evaluated after the resulting antiderivative has been converted back to its original variables, or the original limits can be converted into new limits in terms of the new variables and the antiderivative can be evaluated in the new limits and does not need to be converted back to the original variables.

- **Example:** Find: $_0\int^1$ f(x) = $_0\int^1$ 2x(x² + 1)$^{1/2}$ dx.

First, identify the inside function by looking at which factor is the derivative of the inside function.

Notice that the factor 2x is the derivative of $(x^2 + 1)$. Therefore, choose $u = (x^2 + 1)$.

The derivative of u is: $du/dx = 2x$, or $du = 2x\, dx$

Rearranging gives: $dx = du/2x$

Also, transforming the limits: $u(1) = 2$ and $u(0) = 1$

Next, substitute u and du into the integral:

$$_0\!\int^1 2x(x^2 + 1)^{1/2}\, dx = {}_1\!\int^2 u^{1/2}\, du = (2/3)u^{3/2}\big|_1^2$$
$$= (2/3)(2)^{3/2} - (2/3)(1)^{3/2} = (2/3)[2(2)^{1/2} - 1]$$

This is the final answer because the limits were transformed and the integral was evaluated in the new limits.

Alternatively, the antiderivative in this example can be solved using the substituted integrand, then the result transformed back into the original variables, and evaluated at the original limits:

$$_0\!\int^1 2x(x^2 + 1)^{1/2}\, dx = {}_0\!\int^1 u^{1/2}\, du = (2/3)u^{3/2}$$

Substitute the original expressions back into the evaluated integral, where $u = (x^2 + 1)$:

$$(2/3)u^{3/2} = (2/3)(x^2 + 1)^{3/2}$$

Then evaluate at original limits:

$$(2/3)(x^2 + 1)^{3/2}\big|_0^1 = (2/3)(1^2 + 1)^{3/2} - (2/3)(0^2 + 1)^{3/2}$$
$$= (2/3)(2)^{3/2} - (2/3)(1)^{3/2} = (2/3)[2(2)^{1/2} - 1]$$

Therefore, $_0\!\int^1 2x(x^2 + 1)^{1/2}\, dx = (2/3)[2(2)^{1/2} - 1]$.

Simplifying Integrals Before Using Formal Techniques

• It may be helpful to *simplify an integral* before a more formal technique is employed. Simplifications may include *factoring, substituting trigonometric identities,* or using the *Pythagorean theorem.*

• *Factoring* can also be used to simplify an integral. Examples of factoring include:

$x^2 + (m + n)x + mn \rightarrow$ factors to $\rightarrow (x + m)(x + n)$
$x^2 + 2x + 1 \rightarrow$ factors to $\rightarrow (x + 1)(x + 1)$
$pqx^2 + (pn + qm)x + mn \rightarrow$ factors to $\rightarrow (px + m)(qx + n)$
$x^2 - y^2 \rightarrow$ factors to $\rightarrow (x + y)(x - y)$
$x^2 + 2xy + y^2 \rightarrow$ factors to $\rightarrow (x + y)^2$
$x^2 - 2xy + y^2 \rightarrow$ factors to $\rightarrow (x - y)^2$
$x^2 - 2x + 2 \rightarrow$ factors to $\rightarrow (x - 1)^2 + 1$

Check this last example by working backward:

$(x - 1)(x - 1) + 1 = (x^2 - 2x + 1) + 1 = x^2 - 2x + 2$

• Examples of *trigonometric functions and relations* that can be used when making substitutions include:

$\sin^2 x + \cos^2 x = 1$
$\sin^2 x = 1 - \cos^2 x$
$\cos^2 x = 1 - \sin^2 x$
$1 + \tan^2 x = \sec^2 x$
$\sec^2 x - 1 = \tan^2 x$
$1 + \cot^2 x = \csc^2 x$
$\csc^2 x - 1 = \cot^2 x$
$\tan x = \sin x / \cos x = 1 / \cot x$
$\cot x = \cos x / \sin x = 1 / \tan x = \cos x \csc x$
$\sec x = 1 / \cos x$
$\csc x = 1 / \sin x$
$\sin 2x = 2 \sin x \cos x$
$\cos 2x = \cos^2 x - \sin^2 x = 2 \cos^2 x - 1 = 1 - 2 \sin^2 x$
$\sin(\pi - x) = \sin x$
$\cos(\pi - x) = -\cos x$
$\sin x = \cos(x - \pi/2) = \cos(\pi/2 - x)$
$\cos x = \sin(x + \pi/2) = \sin(\pi/2 - x)$
$\sin(x - y) = \sin x \cos y - \cos x \sin y$
$\cos(x + y) = \cos x \cos y - \sin x \sin y$
$\cos(x - y) = \cos x \cos y + \sin x \sin y$
$\sin x \cos y = (1/2)\sin(x - y) + (1/2)\sin(x + y)$

$\cos x \sin y = (1/2)\sin(x + y) - (1/2)\sin(x - y)$

$\cos x \cos y = (1/2)\cos(x - y) + (1/2)\cos(x + y)$

$\sin x \sin y = (1/2)\cos(x - y) - (1/2)\cos(x + y)$

$\tan(x + y) = (\tan x + \tan y)/(1 - \tan x \tan y)$

$e^{ix} = \cos x + i \sin x$

$e^{-ix} = \cos x - i \sin x$

$e^{i(-x)} = \cos(-x) + i \sin(-x)$

$\cos x = (1/2)(e^{ix} + e^{-ix})$

$\sin x = (1/2i)(e^{ix} - e^{-ix})$

$\cosh x = (1/2)e^{x} + (1/2)e^{-x}$

$\sinh x = (1/2)e^{x} - (1/2)e^{-x}$

$e^{x} = \cosh x + \sinh x$

$e^{-x} = \cosh x - \sinh x$

$\sinh^2 x = (1/2)(\cosh 2x - 1)$

$\cosh^2 x = (1/2)(\cosh 2x + 1)$

$\sinh(x \pm y) = \sinh x \cosh y \pm \cosh x \sinh y$

$\cosh(x \pm y) = \cosh x \cosh y \pm \sinh x \sinh y$

• The following substitutions provide examples of simplifying integrals based on trigonometric identities for the particular selection of u and calculation of du used in the substitution method:

(a.) If an integral contains $[u^2 - a^2]^{1/2}$, substitute for u, $u = a \csc \theta$, then $[u^2 - a^2]^{1/2}$ becomes:

$[a^2\csc^2\theta - a^2]^{1/2} = [a^2(\csc^2\theta - 1)]^{1/2} = [a^2\cot^2\theta]^{1/2} = a \cot \theta$

Note that $(du/d\theta)(a \csc \theta) = -a \csc \theta \cot \theta$.

(b.) Alternatively, if the integral contains $[u^2 - a^2]^{1/2}$, substitute for u, $u = a \sec \theta$, then $[u^2 - a^2]^{1/2}$ becomes:

$[a^2\sec^2\theta - a^2]^{1/2} = [a^2(\sec^2\theta - 1)]^{1/2} = [a^2\tan^2\theta]^{1/2} = a \tan \theta$

Note that $(du/d\theta)(a \sec \theta) = a \sec \theta \tan \theta$.

(c.) If an integral contains $[u^2 + a^2]^{1/2}$, substitute for u, $u = a \tan \theta$, then $[u^2 + a^2]^{1/2}$ becomes:

$$[a^2\tan^2\theta + a^2]^{1/2} = [a^2(\tan^2\theta + 1)]^{1/2} = [a^2\sec^2\theta)]^{1/2}$$
$$= a \sec \theta = a(1/(\cos^2 x))$$

Note that $(du/d\theta)(a \tan \theta) = a \sec^2\theta$.

(d.) If the integral contains $[a^2 - u^2]^{1/2}$, substitute for u, $u = a \sin \theta$, then $[a^2 - u^2]^{1/2}$ becomes:

$$[(a^2 - a^2\sin^2\theta)]^{1/2} = [a^2(1 - \sin^2\theta)]^{1/2} = [a^2\cos^2\theta]^{1/2} = a \cos \theta$$

Note that $(du/d\theta)(a \sin \theta) = a \cos \theta$.

• Examples of *derivatives* to remember when using the substitution method and integration-by-parts:

$(d/dx)\sin x = \cos x$

$(d/dx)\cos x = -\sin x$

$(d/dx)\tan x = 1/(\cos^2 x) = \sec^2 x$

$(d/dx)\cot x = -\csc^2 x$

$(d/dx)\csc x = -\csc x \cot x$

$(d/dx)\sec x = \sec x \tan x$

$(d/dx)\sin^{-1}x = 1/(1 - x^2)^{1/2}$

$(d/dx)\cos^{-1}x = -1/(1 - x^2)^{1/2}$

$(d/dx)\tan^{-1}x = 1/(1 + x^2)$

$(d/dx)\cot^{-1}x = -1/(1 + y^2)$

$(d/dx)\sec^{-1}y = 1/|y|(y^2 - 1)^{1/2}$

$(d/dx)\csc^{-1}y = -1/|y|(y^2 - 1)^{1/2}$

$y = u^n$, $(dy/dx) = nu^{n-1}(du/dx)$, for positive n.

$y = a^u$, $(dy/dx) = a^u(\log a)(du/dx)$

$y = u^v$, $(dy/dx) = vu^{v-1}(du/dx) + u^v(\log u)(dv/dx)$

$y = e^u$, $(dy/dx) = e^u(du/dx)$

$y = \log_a u$, $(dy/dx) = (1/u)\log_a e(du/dx)$

$y = \log u$, $(dy/dx) = (1/u)(du/dx)$, $u \neq 0$.

$y = \cos u$, $(dy/dx) = -\sin u(du/dx)$

$y = \sin u$, $(dy/dx) = \cos u(du/dx)$

$(d/dx)\ln x = 1/x$

• **Example:** Integrate $\int \sin^3 x \, dx$ using substitutions for sine and cosine.

Rearrange:

$$\int \sin^3 x \, dx = \int \sin^2 x \sin x \, dx = \int (1 - \cos^2 x) \sin x \, dx$$

$$= \int (\sin x - \cos^2 x \sin x) \, dx = \int \sin x \, dx - \int \cos^2 x \sin x \, dx$$

Integrate:

$$\int \sin x \, dx = \cos x$$

Then solve $\int \cos^2 x \sin x \, dx$

by substituting $\cos x = u$ and $\sin x \, dx = du$:

$$\int \cos^2 x \sin x \, dx = \int u^2 \, du = u^3/3 + c = (\cos x)^3/3 + c$$

Therefore:

$$\int \sin^3 x \, dx = \int \sin x \, dx - \int \cos^2 x \sin x \, dx = \cos x - (1/3)\cos^3 x + c$$

• Note, the identity $\sin^2 x + \cos^2 x = 1$, is equivalent to $\sin^2 x = 1 - \cos^2 x$ and $\cos^2 x = 1 - \sin^2 x$. These are particularly helpful because they can be applied to sine and cosine raised to other powers. For $\int \sin^2 x \cos x \, dx$ and $\int \cos^2 x \sin x \, dx$, substitutions can be made such as $u = \sin x$, $du = \cos x \, dx$ and $u = \cos x$, $du = -\sin x \, dx$. These substitutions result in integrals in the form $\int u^2 \, du$. Note that other powers of both cosine and sine can be treated similarly and integrated using the substitution method.

3.20 Evaluating Integrals Using Partial Fractions

• This section includes a brief explanation of the method of partial fractions. For a more comprehensive explanation, see a calculus textbook.

• The method of *partial fractions* is applicable when the integrand is a rational fraction such as a quotient of polynomials. In this method, the fraction is separated into a sum of simple fractions that can be easily integrated. The integral of simple fractions often involves the natural logarithm (except when the denominator is raised to a power). When using partial fractions for certain integrals, it may be helpful to remember that $\int 1/x \, dx = \ln|x| + c$, or for a more complicated function, $\int 2/(x + 1) \, dx = 2 \ln|x + 1| + c$.

• Following is a brief explanation of the method of partial fractions and how to use it to solve an integral:

(a.) First verify that the degree of the numerator is smaller than the degree of the denominator. If the degree of the numerator is equal or larger, divide the leading term of the denominator into the leading term of the numerator.

(b.) Next, factor the denominator.

(c.) Then separate the fractions into simpler fractions and insert unknown constants A, B, C, etc., into each numerator. Each factor will generally become a separate fraction. The new numerators will each contain the part of the original denominator (common denominator) that is absent from its new denominator. The constants make the original integral equal to the simpler fractions.

(d.) Then solve for the unknown constants A, B, C, etc., by setting the sum of the new numerators equal to the original numerator.

(e.) Finally, integrate each new fraction resulting in a sum of logarithms.

• A general relation that can be followed for a fraction with a simple binomial in the denominator and a numerator with its degree less than that of the denominator is:

$$\frac{ax + b}{(x - m)(x - n)} \rightarrow \frac{A}{(x - m)} + \frac{B}{(x - n)} \rightarrow \frac{A(x - n) + B(x - m)}{(x - m)(x - n)}$$

where, $(ax + b) = A(x - n) + B(x - m)$.

Using the constants A and B, solve as two integrals:

$$\int [(ax - b)/(x - m)(x - n)] \, dx = \int A/(x - m) \, dx + \int B/(x - n) \, dx$$
$$= A \ln|x - m| + B \ln|x - n| + c$$

• The *partial fractions* method can be demonstrated in this simple example of the integral: $\int 1/(x^2 + 3x + 2) \, dx$

First factor the denominator:

$$\int 1/[(x + 1)(x + 2)] \, dx$$

Translate this integral into two simple integrals:

$$\int A/(x + 1) \, dx + \int B/(x + 2) \, dx$$

where A and B represent constants that make

$1/[(x + 1)(x + 2)]$ equivalent to $[A/(x + 1) + B/(x + 2)]$.

Determine the value of A and B by first combining $[A/(x + 1) + B/(x + 2)]$ back into one fraction with a common denominator (the original denominator) with the proper multiplication steps in the numerators:

$$\frac{A(x+2)+B(x+1)}{(x+1)(x+2)} = \frac{Ax+2A+Bx+B}{(x+1)(x+2)} = \frac{x(A+B)+2A+1B}{(x+1)(x+2)}$$

Then solve for A and B by setting the original numerator "1" equal to the new numerator:

$$1 = x(A + B) + 2A + 1B$$

In order for the expression on the right to equal 1, $A + B$ must equal zero. If this is true, the resulting equation becomes, $1 = 2A + 1B$. This leaves two equations and two unknowns:

$$A + B = 0 \text{ and } 1 = 2A + B$$

A and B can be solved using substitution of the two unknown variables into the two equations. Rearrange $(A + B = 0)$ to isolate A, then substitute the expression for A into $(1 = 2A + B)$:

$$A = -B$$
$$1 = 2(-B) + B = -2B + B = -B$$
$$1 = -B$$

Therefore, $B = -1$.

Substituting B into the $A = -B$:

$$A = -(-1)$$

Therefore, $A = 1$.

Check results by substituting into an original equation:

$$A = -B$$
$$1 = -(-1) = 1$$

Finally solve the simplified (split) integral using the A and B values:

$$\int A/(x + 1)\, dx + \int B/(x + 2)\, dx = \int 1/(x + 1)\, dx - \int 1/(x + 2)\, dx$$
$$= \ln|x + 1| - \ln|x + 2| + c$$

Check this final answer by differentiating:

$$(d/dx)\ln|x + 1| - (d/dx)\ln|x + 2| = 1/(x + 1) - 1/(x + 2)$$

$$\frac{(x+2)-(x+1)}{(x+1)(x+2)} = \frac{x+2-x-1}{(x+1)(x+2)} = \frac{1}{(x+1)(x+2)}$$

which is the original integral.

- A more complicated example using the *partial fractions* method is to solve the integral: $\int (x + 2)/(x^2 + 2x - 3)\, dx$

First factor the denominator:

$$\int (x + 2)/[(x - 1)(x + 3)]\, dx$$

Translate this integral into two simple integrals:

$$\int A/(x - 1)\, dx + \int B/(x + 3)\, dx$$

where A and B represent constants that make

$(x + 2)/[(x - 1)(x + 3)]$ equivalent to $[A/(x - 1) + B/(x + 3)]$.

Determine the value of A and B by first combining $[A/(x - 1) + B/(x + 3)]$ back into one fraction with a common denominator (the original denominator) with the proper multiplication steps in the numerators:

$$\frac{A(x+3)+B(x-1)}{(x-1)(x+3)}$$

To solve for A and B, set the original numerator "(x + 2)" equal to the new numerator:

$$(x + 2) = Ax + 3A + Bx - B$$

Rearranging:

$$2 = Ax + Bx - x + 3A - B$$
$$2 = x(A + B - 1) + 3A - B$$

Set $A + B$ equal to 1 so the x term equals zero. If this is true, the resulting equation becomes $(2 = 3A - B)$.

This leaves two equations and two unknowns:

$$A + B = 1 \text{ and } 2 = 3A - B$$

A and B can be solved using substitution of the two unknown variables into the two equations. Rearrange $(A + B = 1)$ to isolate A, then substitute the expression for A into $(2 = 3A - B)$:

$$A = 1 - B$$
$$2 = 3(1 - B) - B$$
$$2 = 3 - 3B - 1B = 3 - 4B$$
$$-4B = -1$$

Therefore, $B = 1/4$.

Substituting B into the $A = 1 - B$:

$$A = 1 - 1/4 = 3/4$$

Therefore, $A = 3/4$.

Check results by substituting into an original equation:

$$A = 1 - B$$
$$3/4 = 4/4 - 1/4 = 3/4$$

Finally, solve the simplified (split) integral using the A and B values:

$$\int A/(x - 1) \, dx + \int B/(x + 3) \, dx$$
$$= \int (3/4)/(x - 1) \, dx + \int (1/4)/(x + 3) \, dx$$
$$= (3/4) \ln |x - 1| + (1/4) \ln |x + 3| + c$$

Check this final answer by differentiating:

$$(d/dx)(3/4) \ln|x - 1| + (d/dx)(1/4)\ln|x + 3| + c$$

$$= (3/4)(1/(x - 1)) + (1/4)(1/(x + 3))$$

$$= \frac{(3/4)(x+3) + (1/4)(x - 1)}{(x - 1)(x + 3)}$$

$$= \frac{(3x/4) + (9/4) + (x/4) - 1/4}{(x - 1)(x + 3)} = \frac{x + 2}{(x - 1)(x + 3)}$$

which is the original integral.

3.21 Evaluating Integrals Using Tables

• Integral tables are used to solve integrals in forms that do not allow easy application of integration techniques. Integral tables are found in mathematical handbooks, calculus books, online integral tables, and the *CRC Handbook of Chemistry and Physics*.

• Integral tables contain solved integrals in various forms so that an unknown integral can be matched to or translated into the form in the integral table that is identical or most similar to it. If the unknown integral is not identical to a form in the table, a transformation of the integral must be made using substitution. For example, substitute y for ax. Specific substitutions are suggested within integral tables for certain integrals. In general, when making substitutions, a few points that may apply are to make a substitution of the dx terms, to express the limits of the definite integrals in the new dependent variable, and to perform reverse substitution to obtain the answer in terms of the original independent variable.

• In general when using integral tables, identify which type of integral best fits the integral in question. It may be helpful to peruse some integral tables to become familiar with the integrals and substitution suggestions and to read the introductory discussions at the beginning of the tables. Anyone using calculus will often need to look up integrals in the tables. It is worthwhile to become familiar with them. Remember to use laws of logarithms, trigonometric identities, factoring, etc., if necessary to transform an integral into a form that matches a solved integral in the tables. Also, note that sometimes a resulting integral in the tables will have another integral in its result. In these situations, repeat the formula on the resulting integral until there is a constant that results or no further integration is required.

Chapter

4

Series and Approximations

4.1 Sequences, Progressions, and Series

• This section includes sequences, arithmetic, and geometric progressions, and arithmetic and geometric series.

• A *sequence* is a set of numbers called *terms*, which are arranged in a succession in which there is a relationship or rule between each successive number. A sequence can be finite or infinite. A *finite sequence* has a last term and an *infinite sequence* has no last term.

• The following is an example of a finite sequence:

{3, 6, 9, 12, 15, 18}

In this sequence each number has a value of 3 more than the preceding number.

• The following is an example of an infinite sequence describing the function $f(x) = 1/x$:

{1/1, 1/2, 1/3, 1/4, 1/5, 1/6,...}

where the *domain set* is x = {1, 2, 3, 4, 5, 6,...} and the *range set* is $f(x)$ = {1/1, 1/2, 1/3, 1/4, 1/5, 1/6,...}.

• An *arithmetic progression* is a sequence in which the difference between successive terms is a fixed number and each term is obtained by adding a fixed amount to the term before it. This fixed amount is called the *common difference*. Arithmetic progressions can be represented by first-degree polynomial expressions. For example, the expression $(n + 1)$ can represent an arithmetic progression.

• The sequence {3, 6, 9, 12, 15, 18} is an arithmetic progression represented by $(n + 3)$.

• A *finite arithmetic progression* can be expressed as:

a, a + d, a + 2d, a + 3d, a + 4d, a + 5d,..., a + (n – 1)d

where a is the first term, d is the fixed difference between each term, and $(a + (n – 1)d)$ is the last or "nth" term. Each term in this progression can be written as:

$$n = 1, \quad a_1 = a + (1 – 1)d = a$$
$$n = 2, \quad a_2 = a + (2 – 1)d = a + d$$
$$n = 3, \quad a_3 = a + (3 – 1)d = a + 2d$$
$$n = 4, \quad a_4 = a + (4 – 1)d = a + 3d$$
$$n = 5, \quad a_5 = a + (5 – 1)d = a + 4d$$

and so on.

• For example, in the arithmetic progression {3, 6, 9, 12, 15, 18}, a = 3 and d = 3. Therefore,

$$\text{for } n = 1, \quad a_1 = 3,$$
$$\text{for } n = 2, \quad a_2 = 6,$$
$$\text{for } n = 3, \quad a_3 = 9,$$

and so on.

• A *geometric progression* is a sequence in which the *ratio* of successive terms is a fixed number, and each term is obtained by multiplying a fixed amount to the term before it. This fixed amount is called the *common ratio*.

• Terms in a geometric progression can be represented as:

$a, ar, ar^2, ar^3, ar^4, ar^5, ..., ar^{n-1}$

where a is the first term, ar^{n-1} is the last term and the ratio of successive terms is given by r such that:

$ar/a = r, ar^2/ar = r, ar^3/ar^2 = r$, etc.

• For example, in the geometric progression $\{2, 4, 8, 16, 32,...\}$, if $a = 2$ and $r = 2$, the geometric progression can be expressed as:

$2, 2(2), 2(2)^2, 2(2)^3, 2(2)^4, ..., 2(2)^{n-1}$

• A *series* is the *sum* of the terms in a progression or sequence. An *arithmetic series* is the sum of the terms in an arithmetic progression. A *geometric series* is the sum of the terms in a geometric progression.

• The notation used to express a series is *sigma notation*. The sigma notation that represents an *arithmetic series* is:

$$\sum_{n=1}^{m} a_n$$

where a_n is the sequence function, m is the last term that is added, and n is the nth term.

The sum of the first three terms in sequence a_n from $n = 1$ to $n = 3$, can be represented using sigma notation:

$$\sum_{n=1}^{3} a_n = a_1 + a_2 + a_3$$

• For example, in the arithmetic progression $\{3, 6, 9, 12,...\}$ the sum of the first three terms is the *arithmetic series*:

$$\sum_{n=1}^{3} a_n = 3 + 6 + 9 = 18$$

• An arithmetic series can be calculated by determining the sum of the terms in an arithmetic progression using the *formula* $(m/2)(a_1 + a_m)$ where m represents the last term added.

• For example, applying this formula to the arithmetic progression {3, 6, 9} results in: $(3/2)(3 + 9) = (3/2)(12) = 18$

• A *geometric series* can also be represented using sigma notation as follows:

$$\sum_{n=1}^{m} ar^{n-1} = a + ar + ar^2 + ar^3 + ar^4 + ar^5 + ... + ar^{n-1}$$

where a is the first term and $a \neq 0$, r is the ratio between successive terms, m is the index of the last term added, n is the nth term, and ar^{n-1} is the last term.

• For example, in the geometric progression {2, 4, 8, 16, 32,...} the sum of the first three terms is the *geometric series*:

$$\sum_{n=1}^{3} ar^{n-1} = 2 + 4 + 8 = 14$$

• A geometric series can be calculated by determining the sum of the terms in the geometric progression using the *formula* $[(a)(1 - r^m)/(1 - r)]$ where m represents the index of the last term added, and r is the ratio.

• For example, applying this formula to the geometric progression {2, 4, 8} results in: $2(1 - 2^3)/(1 - 2) = 14$

• In an *infinite geometric series*, m approaches infinity. As m approaches infinity, the formula for the series becomes:

$$\lim_{m \to \infty} [a(1 - r^m)/(1 - r)]$$

If $|r| < 1$ and $m \to \infty$, then r^m approaches zero and the sum of the infinite geometric series becomes $a/(1 - r)$.

• Because a series can be differentiated, multiplied, added to, etc., it is sometimes written in terms of the variable x rather than r:

$$a + ax + ax^2 + ax^3 + ax^4 + ... + ax^{n-1} = a(1 - x^n)/(1 - x)$$

4.2 Infinite Series and Tests for Convergence

• This section includes infinite series and convergence, convergence of a geometric series, tests for convergence including the Comparison Test, the Ratio Test, tests for series with positive and negative terms, the Integral Test, and the Root Test.

• A *series is infinite* if there are an infinite number of terms in the progression or sequence that define the series. If the progression or sequence has an infinite number of terms, then the sum cannot be calculated exactly. However, under certain conditions the sum can be estimated.

• Conditions that determine whether the sum of an infinite series can be estimated include the following:

(a.) If an infinite series has a *limit* it will *converge* and its sum can be estimated. If as the terms in an infinite series are added where with each additional term added the sum approaches a number, then the series has a limit and converges and the sum can be estimated.

(b.) If an infinite series has no limit it will *diverge* and the sum cannot be estimated. If as each additional term is added the sum approaches infinity, then the series has no limit and diverges and the sum cannot be estimated.

(c.) A condition for convergence for infinite series:

$$\sum_{n=1} a_n$$

is that a_n must approach zero as n approaches infinity.

Although this condition must occur for a series to converge, there are cases where this condition is true but the series still diverges.

• To *estimate an infinite series* it must be determined whether the series has a limit and converges and what happens to the sum as the number of terms approaches infinity. For example, consider the infinite series describing the sum of a_n from n = 1 to n = ∞:

$$\sum_{n=1} a_n$$

If this series has a limit and converges to L, it becomes:

$$\lim_{n \to \infty} \sum_{n=1} a_n = L$$

• The *geometric series*: $a + ar + ar^2 + ar^3 + ar^4 + \ldots + ar^{n-1}$ converges when $|r| < 1$ and diverges when $|r| \geq 1$.

This geometric series can be expressed as:

$$\sum_{n=1}^{m} ar^{n-1} = a + ar + ar^2 + ar^3 + ar^4 + \ldots + ar^{n-1}$$

where the sum of the first m terms is represented by the formula:
$a(1 - r^m)/(1 - r) = a/(1 - r) - (ar^m)/(1 - r)$

As m approaches infinity the formula becomes:

$$\lim_{m \to \infty} [\, a/(1 - r) - (ar^m)/(1 - r)]$$

If $|r| < 1$ and $m \to \infty$, then $r^m \to 0$ and the formula becomes:

$a/(1 - r)$

Therefore, as $m \to \infty$, if $|r| < 1$, the series converges and if $|r| \geq 1$, the series diverges.

• To determine whether an infinite series will converge, there are a variety of *tests for convergence* that may be used. These tests include *the Comparison Test*, the *Ratio Test*, tests for series with positive and negative terms, the *Integral Test,* and the *Root Test*.

The Comparison Test for Convergence

• The *Comparison Test* can be applied to infinite series with positive terms. A series is *convergent* if each term is less than or equal to each corresponding term in a series that is known to be convergent. Conversely, if each term in an unknown series is greater than or equal to each corresponding term in a known *divergent* series, then the unknown series is also divergent.

• An example of a known *convergent series* that is used in a Comparison Test is the *P Series*:

$$1 + 1/2^P + 1/3^P + \ldots + 1/n^P + \ldots$$

This series converges when $P > 1$ and diverges when $P \leq 1$.

• A *divergent series* that is used in the Comparison Test is:
$$1 + 1 + 1 + 1 + \ldots$$

As the number of terms approaches infinity, the sum of the terms approaches infinity and the series diverges.

• **Example:** Will series U converge?

$$U = 1 + 1/2 + 1/3 + \ldots + 1/n + \ldots$$

(This is called the Harmonic Series.)

Compare with the known series K, which diverges as more terms are added:

$$K = 1 + 1/2 + 1/2 + 1/2 + \ldots$$

To compare series K to series U, rewrite series K as follows and compare the two series term by term:

$$K = 1 + (1/2) + (1/4 + 1/4) + (1/8 + 1/8 + 1/8 + 1/8) + \ldots$$
$$U = 1 + 1/2 + 1/3 + 1/4 + 1/5 + 1/6 + 1/7 + 1/8 + \ldots$$

Many of the terms in U are greater than the corresponding terms in K. For example, the third terms $(1/3 > 1/4)$ and the fifth terms $(1/5 > 1/8)$. Therefore, because K diverges U must also diverge.

This example is interesting because in the Harmonic Series the value of the terms do approach zero, which is a necessary criterion for convergence but does not guarantee it. By applying the Comparison Test with a series that is known to diverge, it is clear that the Harmonic Series diverges.

The Ratio Test for Convergence

• The *Ratio Test* for convergence can be applied to a series of positive terms and to a series containing positive and negative terms.

- To apply the Ratio Test for the series of positive terms

$$a_1 + a_2 + a_3 + \ldots a_n + \ldots,$$

find the ratio r of successive terms: $r = (a_{n+1}/a_n)$

To determine r, take the limit as n→∞: $r = \lim_{n\to\infty}(a_{n+1}/a_n)$

    If $r < 1$, the series will converge.

    If $r > 1$, the series will diverge.

    If $r = 1$, test does not indicate convergence or divergence.

- A *generalized Ratio Test* can be applied to the power series:

$$\sum_{n=0}^{\infty} a_n x^n$$

$$\lim_{n\to\infty} \frac{|a_{n+1}x^{n+1}|}{|a_n x^n|} = \lim_{n\to\infty} |x|\frac{|a_{n+1}|}{|a_n|}$$

If $|x| < \lim_{n\to\infty} \dfrac{|a_{n+1}|}{|a_n|}$, the series converges.

If $|x| > \lim_{n\to\infty} \dfrac{|a_{n+1}|}{|a_n|}$, the series diverges.

If $|x| = \lim_{n\to\infty} \dfrac{|a_{n+1}|}{|a_n|}$, the series may or may not converge.

Note that the set of values of x for which the series is convergent is called the *interval of convergence*.

- The Ratio Test can be applied to evaluate convergence of series containing *positive* and *negative* terms. To apply the Ratio Test to an *alternating series*, take the limit as n→∞ for the ratio of the absolute value of successive terms:

$$\lim_{n\to\infty} \frac{|a_{n+1}|}{|a_n|} = r$$

    If $r < 1$, the series will converge.

    If $r > 1$, the series will diverge.

    If $r = 1$, test does not indicate convergence or divergence.

Tests for Series with Positive and Negative Terms

• A *series with positive and negative terms* converges if the corresponding series of *absolute values* of the terms converges. If series $|S|$ converges, then series S will converge.

Series $|S|$ is given by:

$$|S| = |a_1| + |a_2| + |a_3| + |a_4| + ... |a_n| + ...$$

Corresponding series S is given by:

$$S = a_1 + a_2 + a_3 + a_4 + ... a_n + ...$$

where a_n can be positive or negative.

• A series with positive and negative terms may converge and is called *conditionally convergent* even though its corresponding series of absolute values diverges. For example:

$$1 - 1/2 + 1/3 - 1/4 + 1/5 - ... \text{ converges conditionally.}$$
$$1 + 1/2 + 1/3 + 1/4 + 1/5 ... \text{ diverges.}$$

• In an *alternating series* the signs of the terms alternate positive and negative:

$$a_1 + a_2 - a_3 + a_4 - a_5 + a_6 - a_7 + ... a_n + ...$$

The alternating series will converge if the following conditions are true from some point in the series: $a_n \geq a_{n+1}$ for all values of n, each a is positive and $\lim_{n \to \infty}[a_n] = 0$.

Integral Test for Convergence

• The *Integral Test* can be applied to a decreasing series of positive terms in which $a_{n+1} < a_n$ for all successive terms. To apply the Integral Test to a series, integrate the function representing the series. If the integral of the series exists and therefore converges, then the series also converges.

• Consider the decreasing series:

$$\sum_{n=1}^{\infty} a_n$$

where a_n represents $f(x)$. If $f(x)$ is a positive continuous function and $\int_1^{\infty} f(x) \, dx$ exists and converges, then the series also converges.

• For example, to apply the Integral Test to the series represented by $f(x) = 1/x$, integrate between 1 and ∞:

$$\int_1^\infty (1/x)dx = \ln x \Big|_1^\infty = \ln \infty - \ln 1 = \infty$$

The integral of $1/x$ is $\ln x$, (ln is the natural logarithm.) Because the integral from 1 to ∞ is infinity and does not exist, it diverges. Therefore, the series diverges.

Root Test for Convergence

• The *Root Test* can be applied to series

$a_1 + a_2 + a_3 + a_4 + ... a_n + ...$, such that: $\lim_{n\to\infty} \sqrt[n]{|a_n|} = r$

If $r < 1$, the series will converge.

If $r > 1$, the series will diverge.

If $r = 1$, test does not indicate convergence or divergence.

4.3 Expanding Functions Into Series, the Power Series, Taylor Series, Maclaurin Series, and the Binomial Expansion

• This section includes expanding functions into series, the power series, the Maclaurin and Taylor series, and the binomial expansion.

• When a *function* is written in the form of an *infinite series*, it is said to be *"expanded"* in an infinite series.

• In general, a function $f(x)$ expanded in an infinite power series is written:

$$f(x) = \sum_{n=0}^{\infty} a_n(x-a)^n =$$

$$a_0 + a_1(x-a) + a_2(x-a)^2 + a_3(x-a)^3 + a_4(x-a)^4 + ...a_n(x-a)^n...$$

Or when $a = 0$:

$$f(x) = \sum_{n=0}^{\infty} a_n x^n = a_0 + a_1 x + a_2 x^2 + a_3 x^3 + a_4 x^4 + a_5 x^5 + ...+ a_n x^n...$$

Where $a_0, a_1,...a_n$ represent constant coefficients, x is a variable, and a is a constant called the *center of the series*.

The power series in x converges if x = 0 or it converges for all x at a radius of convergence r such that if $|x| < r$ it converges and if $|x| > r$ it diverges.

The power series in (x – a) converges if x = a. If a = 0, the power series in x results.

The function f(x) has the following properties of a polynomial: (a.) It is continuous within the interval of convergence (there is no break in its graph); (b.) in series form, the function can be added, subtracted, multiplied, or divided term by term; and (c.) if f(x) is differentiable, then the series can be differentiated term by term.

• There is a positive number r called the *radius of convergence* where the power series converges if $|x - a| < r$ and diverges if $|x - a| > r$. The number r can represent a circle of convergence where the series may or may not converge for all points on the circle of convergence. The inequality $|x - a| < r$ is sometimes called the *interval of convergence*.

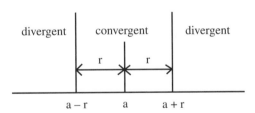

• Two common series representing expansions of functions are the *Maclaurin series* and the *Taylor series*. Expanding functions into these series can be applied to approximating functions including linear and quadratic approximations, approximating solutions to differential equations, and estimating numerical values such as constructing tables of exponential, logarithmic, and trigonometric functions.

• Representing a function in a Taylor series or a Maclaurin series involves determining the *coefficients* a_0, a_1,...a_n of the series. The coefficients can be found by *differentiation* providing the function has all its derivatives. Obtaining all the derivatives of a function can be tedious, so other methods including substitution and integration are employed.

- If function $f(x)$ is expanded in a *power series*, the result is:

$$f(x) = \sum_{n=0}^{\infty} a_n(x-a)^n =$$

$$a_0 + a_1(x-a) + a_2(x-a)^2 + a_3(x-a)^3 + \ldots a_n(x-a)^n + \ldots$$

In the special case of $a = 0$, the result is:

$$f(x) = a_0 + a_1x + a_2x^2 + a_3x^3 + a_4x^4 + \ldots a_nx^n + \ldots$$

where, $f(a) = a_0$.

Determine the coefficients at $x = a = 0$.

First take the first derivative of each term:

$$f'(x) = a_1 + 2a_2x^1 + 3a_3x^2 + 4a_4x^3 + \ldots na_nx^{n-1} + \ldots$$

where, $f'(a) = f'(0) = a_1$.

Take the second derivative of each term:

$$f''(x) = 2a_2 + (2)3a_3x + \ldots n(n-1)a_nx^{n-2} + \ldots$$

where, $f''(a) = f''(0) = 2a_2$.

Take the third derivative of each term:

$$f'''(x) = 2(3)a_3 + (2)(3)4a_4x + \ldots n(n-1)(n-2)a_nx^{n-3} + \ldots$$

where $f'''(a) = f'''(0) = 6a_3$.

Take the nth derivative of each term:

$$f^{(n)}(x) = n!a_n + (n+1)!a_{n+1}x + \ldots$$

(Remember "!" represents *factorial*.)

If the coefficients are determined at $x = a = 0$:

$$a_0 = f(0)$$
$$a_1 = f'(0)$$
$$a_2 = f''(0)/2$$
$$a_3 = f'''(0)/6$$
$$a_n = f^{(n)}(0)/n!$$

Therefore, the expansion of $f(x)$ about $x = a = 0$ is:

$$f(x) = [f(0)] + [f'(0)]x + [f''(0)/2!]x^2 + [f'''(0)/3!]x^3 + \ldots [f^{(n)}(0)/n!]x^n \ldots$$

$$= a_0 + a_1x + a_2x^2 + a_3x^3 + a_4x^4 + \ldots a_nx^n \ldots = \sum_{n=0}^{\infty} \frac{f^{(n)}(0)}{n!}x^n$$

This is known as the *Maclaurin series* or the Taylor series for $f(x)$ expanded about the point $x = 0$.

• In the *Taylor series*, the function is generally expanded about some point a rather than zero. For the function $f(x)$:

$$f(x) = a_0 + a_1(x-a) + a_2(x-a)^2 + a_3(x-a)^3 + \ldots a_n(x-a)^n + \ldots$$

The coefficients a_n are computed by repeated differentiation as with the Maclaurin series. The resulting Taylor series for $f(x)$ is:

$$f(x) = [f(a)] + [f'(a)](x-a) + [f''(a)/2!](x-a)^2 + [f'''(a)/3!](x-a)^3$$

$$+ \ldots + [f^{(n)}(a)/n!](x-a)^n \ldots = \sum_{n=0}^{\infty} \frac{f^{(n)}(a)}{n!}(x-a)^n$$

This is the *Taylor series*, which is expanded about point $x = a$. If $a = 0$, the Taylor Series becomes the Maclaurin Series.

• **Example:** To write the Taylor series expansion of $\ln x$ near 1, determine the coefficients and substitute $a = 1$ into the above expression.

$$f(x) = \ln x \rightarrow f(a=1) = 0$$
$$f'(x) = 1/x \rightarrow f'(a=1) = 1$$
$$f''(x) = -1/x^2 \rightarrow f''(a=1) = -1$$
$$f'''(x) = 2/x^3 \rightarrow f'''(a=1) = 2$$
$$f^{(4)}(x) = -6/x^4 \rightarrow f^{(4)}(a=1) = -6$$

and so on.

The expansion of $f(x) = \ln x$ about the point $a = 1$ is:

$$\ln x = 0 + (x-1) - (x-1)^2/2! + 2(x-1)^3/3! - 6(x-1)^4/4! + \ldots$$
$$= (x-1) - (x-1)^2/2 + (x-1)^3/3 - (x-1)^4/4 + \ldots + (-1)^{n-1}(x-1)^n/n$$

• Taylor series are generally good approximations when x is near a. A series will often converge at different locations depending upon the value of x. For example, in the Taylor series expansion of ln x, the values of x where the series converges are between x = 0 and x = 2. Therefore, the *interval of convergence* for ln x is 0 < x < 2. The ln x series will converge faster near x = 1 than at the extremes 0 and 2.

• **Example:** The *exponential function e^x* can be computed using the Taylor or Maclaurin expansions:

The Maclaurin expansion of e^x is:

$$e^x = 1 + x + x^2/2! + x^3/3! + x^4/4! + ... + x^n/n! + ...$$

For x = 1, this becomes:

$$e^x = 1 + 1 + 1/2! + 1/3! + 1/4! + ... + 1/n! + ...$$
$$= 1 + 1 + 0.5 + 0.166667 + 0.041667 + 0.008333$$
$$+ 0.001389 + 0.000198 + ... = 2.718254$$

Therefore, for x=1, e^1 is approximately equal to 2.718254.

For the *Taylor expansion of e^x* near a = x = 1, all the derivatives are e:

$$e^x = e + e(x-1) + e(x-1)^2/2! + e(x-1)^3/3! + ...$$

• **Example:** The Taylor series of $\exp\{-x^2\}$ about zero can be found by substituting $u = -x^2$ rather than differentiating $\exp\{-x^2\}$ directly.

$$e^u = 1 + u + u^2/2! + u^3/3! + u^4/4! + ...$$

Substituting back:

$$= 1 + \{-x^2\} + \{-x^2\}^2/2! + \{-x^2\}^3/3! + \{-x^2\}^4/4! + ...$$

Therefore, $\exp\{-x^2\} = 1 - x^2 + x^4/2! - x^6/3! + x^8/4! + ...$

• **Example:** *Trigonometric functions can be expanded* and computed for selected values. The expansions of sine and cosine are:

$$\sin x = x - x^3/3! + x^5/5! - x^7/7! + ... + (-1)^{n-1}x^{2n-1}/(2n-1)! + ...$$
$$\cos x = 1 - x^2/2! + x^4/4! - x^6/6! + ... + (-1)^{n-1}x^{2n-2}/(2n-2)! + ...$$

- The series for e^x, sin x, and cos x all have $x^n/n!$ terms where the factorials lead to convergence for all x. Also, term by term *differentiation* of series e^x yields e^x, and term by term differentiation of series sin x yields series cos x.

- **Example:** The *geometric series* is also a Taylor series obtained by taking derivatives. The geometric series is an expansion of $f(x) = 1/(1 - x)$ near zero and is given by:

$$1 + x + x^2 + x^3 + x^4 + ... = 1/(1 - x)$$

This series converges for $|x| < 1$, and at x = 1 the point lies on the *circle of convergence*.

- **Example:** The series for $e^{i\theta}$:

$$e^{i\theta} = 1 + i\theta + (i\theta)^2/2! + (i\theta)^3/3! + ...$$

can be shown to equal $(\cos \theta + i \sin \theta)$ as follows:

$$\cos \theta + i \sin \theta = [1 - \theta^2/2! + \theta^4/4! - \theta^6/6!...] + i[\theta - \theta^3/3! + \theta^5/5! -...]$$
$$= 1 + i\theta - \theta^2/2! - i\theta^3/3! + \theta^4/4! + i\theta^5/5! - \theta^6/6!...$$

Substitute i^2 for -1, i^3 for $-i$, i^4 for 1, i^5 for i, etc:

$$\cos \theta + i \sin \theta = 1 + i\theta + (i\theta)^2/2! + (i\theta)^3/3! + (i\theta)^4/4! + (i\theta)^5/5! ...$$
$$= e^{i\theta} = [1 - \theta^2/2! + \theta^4/4! - \theta^6/6!...] + i[\theta - \theta^3/3! + \theta^5/5! -...] =$$
$$e^{i\theta} = \cos \theta + i \sin \theta$$

which is *Euler's formula*, where the real part is x = cos θ and the imaginary part is y = sin θ, and the x and y coordinates designate $e^{i\theta}$ on the complex plane with a radius of 1 (because $\cos^2\theta + \sin^2\theta = 1$).

y = r sinθ

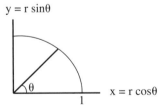

x = r cosθ

1

where $re^{i\theta}$ = r cos θ + ir sin θ = x + iy.

• A *binomial expression* (a + b) can be expanded into polynomial form called a *binomial expansion*. To expand (a + b) into (a + b)^n, first consider the expansions for

(a + b)^2, (a + b)^3, and (a + b)^4:

(a + b)^2 = (a + b)(a + b) = a^2 + ab + ab + b^2 = a^2 + 2ab + b^2

(a + b)^3 = (a + b)(a + b)(a + b) = a^3 + 3a^{2}b + 3ab^2 + b^3

(a + b)^4 = (a + b)(a + b)(a + b)(a + b)

= a^4 + 4a^{3}b + 6a^{2}b^2 + 4ab^3 + b^4

These expansions are obtained by multiplying the first two binomials, then multiplying each successive binomial with the preceding polynomial.

For the expansion of (a + b)^n where n is a positive integer, the *Binomial Theorem* is applied as follows:

(a + b)^n = a^n + na^{n-1}b + [n(n–1)/(2)(1)] a^{n-2}b^2

+ [n(n–1)(n–2)/(3)(2)(1)] a^{n-3}b^3 +...

+ [n(n–1)(n–2)...(n–r+2)/(r–1)!] (a^{n-r+1}b^{r-1}) +...+ b^n

The rth term is given by:

[n(n–1)(n–2)...(n–r+2)/(r–1)!] (a^{n-r+1}b^{r-1})

where r represents some integer between 1 and n.

• The Taylor series for (1 + x)^p is called a binomial series given by:

(1 + x)^p = 1 + px + [p(p–1)/2!]x^2 + [p(p–1)(p–2)/3!]x^3 +...

where p is a positive integer.

This series converges for $|x| < 1$.

Chapter

5

Vectors, Matrices, Curves, Surfaces, and Motion

5.1 Introduction to Vectors

• This section includes definitions, notation, types of vectors including displacement, velocity, zero, unit, equivalent, position, and addition and subtraction of vectors.

• *Scalars* are quantities that represent magnitude and can be described by one number that is positive, negative, or zero. Scalars can be compared with each other when they have the same physical dimensions or units. Examples of scalars include temperature, work, density, and mass.

• A *vector* represents a quantity that is described by both a numerical value for *magnitude* (or *length*) and a *direction*. A vector is depicted as a line segment with an initial point and a terminal point that has an arrow pointing in the direction of the terminal point. Examples of vectors include displacement, velocity, acceleration, electric field strength, force, and moment of force.

• A *displacement vector* represents the change or displacement between two points in a coordinate system. The *length* of a displacement vector is the distance between the two points and the *direction* of a displacement vector is the direction it is pointing.

• A *velocity vector* describes an object in motion and has a magnitude representing the speed of the object and a direction representing the direction of motion.

• *Notation for a vector* includes boldface letters **A**, **a**, **B**, **b**, etc., or one or two letters with an arrow $\vec{A}$, $\vec{B}$, $\vec{a}$, $\overrightarrow{AB}$, etc. Vectors can be written in terms of their components on a coordinate system. For example:

$$\mathbf{A} = a_1\mathbf{i} + a_2\mathbf{j} + a_3\mathbf{k} = [a_1, a_2, a_3]$$

where a_1, a_2, a_3 are the *components*.

Vectors can also be written in the form of *column vectors* and *row vectors*:

$$\mathbf{A} = \begin{bmatrix} a_1 \\ a_2 \end{bmatrix}, \quad \mathbf{B} = [b_1 \quad b_2], \quad \mathbf{v} = \begin{bmatrix} v_1 \\ v_2 \end{bmatrix}, \quad \mathbf{r} = [r_1 \quad r_2]$$

where a_1 and a_2 are components of **A**, b_1 and b_2 are components of **B**, v_1 and v_2 are components of **v**, and r_1 and r_2 are components of **r**. *Unit vectors i, j, and k* have directions pointing parallel to the axes of a coordinate system.

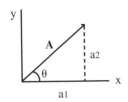

• The *magnitude (or length)* of a vector is denoted with vertical bars as used with absolute value. For example, the following represent magnitudes: $|\mathbf{A}|$, $|\mathbf{B}|$, $|\overrightarrow{AB}|$. Sometimes double bars are used to represent magnitude: $||\mathbf{A}||$, $||\mathbf{B}||$.

• If vector $\mathbf{A} = \overrightarrow{ab}$ and points from a to b and vector $\mathbf{B} = \overrightarrow{ba}$ and points from b to a, then $\mathbf{A} = -\mathbf{B}$.

• When a vector is changed from its column format to a row format or vice versa, it is called a *transposition* and indicated by "T".

Therefore, if $\mathbf{v} = \begin{bmatrix} v_1 \\ v_2 \end{bmatrix}$, then $\mathbf{v}^T = [v_1 \quad v_2]$.

Similarly, if $\mathbf{v} = [v_1 \quad v_2]$, then $\mathbf{v}^T = \begin{bmatrix} v_1 \\ v_2 \end{bmatrix}$.

• The *zero vector* **0** has a length (or magnitude) of zero and no direction. Its initial and terminal points coincide.

• A *unit vector* **u** has a length (or magnitude) of one. If unit vector **u** is pointing in the direction of vector **A** and **A** is not a zero vector, then $\mathbf{u} = \mathbf{A} / |\mathbf{A}|$.

• Vectors that point in the same direction and have the same length are *equivalent vectors* even if they are not in the same location. A vector can be relocated and still be considered the same vector as long as its length and direction remain the same.

• A vector with its initial point at the origin of a *coordinate system* is called a *position vector*. A position vector is defined according to the location or coordinates of its terminal point. For example, if its terminal point is at B then vector $\overrightarrow{AB}$ is a position vector of point B. A position vector represents the position of a point with respect to the origin and a *displacement vector* represents the change or displacement between two points in a coordinate system.

• A vector is often written in terms of its *components*, which are defined by its directions along the XYZ axes of a coordinate system. Generally, *i, j, and k* are *unit vectors* with magnitudes of one and directions pointing parallel to the XYZ axes respectively in a rectangular coordinate system.

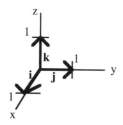

• Vector **A** can be written using the i, j, k unit vectors as: $\mathbf{A} = a_1\mathbf{i} + a_2\mathbf{j} + a_3\mathbf{k}$ where a_1, a_2, and a_3 are scalar quantities and $a_1\mathbf{i}$, $a_2\mathbf{j}$, and $a_3\mathbf{k}$ are the components of **A**. The *magnitude (or length)* of **A** is given by:

$$|\mathbf{A}| = \sqrt{a_1^2 + a_2^2 + a_3^2}$$

• If a *position vector* has its starting point at the origin and its terminal point at point P = (5,6), then in two dimensions vector **A** is written:

$$\mathbf{A} = 5\mathbf{i} + 6\mathbf{j}$$

It has length $|\mathbf{A}| = \sqrt{(5)^2 + (6)^2} = \sqrt{25 + 36} = \sqrt{61}$ and is depicted as:

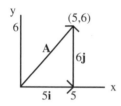

• Any vector **A** can be multiplied by a constant c such that: $c\mathbf{A} = ca_1\mathbf{i} + ca_2\mathbf{j} + ca_3\mathbf{k}$

• The *zero vector* having zero length can be written in terms of i, j, k:

$$\mathbf{0} = 0\mathbf{i} + 0\mathbf{j} + 0\mathbf{k}$$

• Unit vectors i, j, k can be represented as *column vectors*:

$$\mathbf{i} = \begin{bmatrix} 1 \\ 0 \\ 0 \end{bmatrix}, \mathbf{j} = \begin{bmatrix} 0 \\ 1 \\ 0 \end{bmatrix}, \mathbf{k} = \begin{bmatrix} 0 \\ 0 \\ 1 \end{bmatrix}$$

Vector **A** can be written in column vector format:

$$\mathbf{A} = 5\mathbf{i} + 3\mathbf{j} - 6\mathbf{k} = \begin{bmatrix} 5 \\ 3 \\ -6 \end{bmatrix}$$

• Vectors can be characterized and expressed in more than three dimensions or components. For example:

$$\mathbf{A} + \mathbf{B} = (a_1, a_2, a_3, a_4) + (b_1, b_2, b_3, b_4)$$

- The *direction of a vector* in a coordinate system is represented by the angle it makes with the positive X-axis. For example, the direction of vector **A** can be written in terms of the angle θ that it makes with the positive X-axis. Vector $\mathbf{A} = a_1\mathbf{i} + a_2\mathbf{j}$ makes an angle $\theta = \tan^{-1}(a_2/a_1)$ with the X-axis and can be written as:

$$\mathbf{A} = \mathbf{i}\,|\mathbf{A}|\,\cos\theta + \mathbf{j}\,|\mathbf{A}|\,\sin\theta$$

where $a_1 = |\mathbf{A}|\,\cos\theta$ and $a_2 = |\mathbf{A}|\,\sin\theta$.

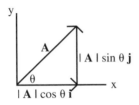

A *unit vector* **u** for vector **A** can be written as:

$$\mathbf{u} = \frac{\mathbf{A}}{|\mathbf{A}|} = \mathbf{i}\cos\theta + \mathbf{j}\sin\theta = \begin{bmatrix} \cos\theta \\ \sin\theta \end{bmatrix}$$

Also, $|\mathbf{u}|^2 = \cos^2\theta + \sin^2\theta = 1$

- A vector **v** divided by its *length* $|\mathbf{v}|$ results in a *unit vector* pointing in the *direction of the vector*. The direction of **v** is $\mathbf{u} = \mathbf{v}/|\mathbf{v}|$ and its length is $|\mathbf{v}|$. Therefore, length multiplied by direction gives **v** as $\mathbf{u}|\mathbf{v}| = \mathbf{v}$.

- In three dimensions the components of a unit vector **U** are called "*direction cosines*" and have angles α, β, and γ with the X-, Y-, Z-axes, respectively. In three dimensions:

$$\mathbf{U} = \mathbf{i}\cos\alpha + \mathbf{j}\cos\beta + \mathbf{k}\cos\gamma$$

and $\cos^2\alpha + \cos^2\beta + \cos^2\gamma = 1$

Addition and Subtraction

- Two *vectors can be added or subtracted* if they have the same dimensions by adding or subtracting the corresponding components (or elements). For example, a two-dimensional vector can be added to another two-dimensional vector, however a two-dimensional vector cannot be added to a three-dimensional vector.

• The *sum of two vectors* can be depicted by positioning the vector such that the initial point of the second vector is at the terminal point of the first vector. The sum of the two vectors is a third vector with its initial point at the initial point of the first vector and its final point at the final point of the second vector. In other words, the sum of two vectors **a** and **b** is the combined displacement from applying vector **a** then applying vector **b**. Consider the figure below depicting the following two examples of adding vectors **a** and **b**:

For example, to add vectors **a** and **b** in the first illustration, place the initial point of **b** at the final point of **a**. The sum is the vector joining the initial point of **a** to the final point of **b**, or vector **c**.

In the second illustration, the initial point of **b** is already at the final point of **a**. The sum is the vector joining the initial point of **a** to the final point of **b**, which results in vector **c**.

Remember that the starting point of a vector can be moved as long as its length and direction stay the same. Also note that the sum is also the diagonal of a parallelogram that can be constructed on **a** and **b**.

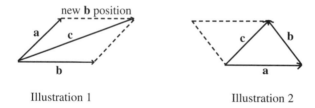

Illustration 1   Illustration 2

Both figures represent **a** + **b** = **c**.

• *Subtraction of two vectors* is equivalent to addition of the first vector with the negative of the second vector. The *negative of a vector* is a vector with the same length but pointing in the opposite direction.

• To subtract two vectors, reverse the direction of the second vector, then add the first vector with the negative of the second vector by positioning the vectors so that the initial point of the (negative) second vector is at the final point of the first vector. The sum of two vectors will be a third vector with its initial point at the initial point of the first vector and its final point at the final point of the second (negative) vector. This figure represents **a** − **b** = **c**:

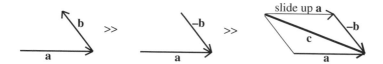

In a second example of vector subtraction, subtract two vectors **a** – **b** = **c**, where **a** – **b** can be represented using the negative of **b**, then slide –**b** up to place initial point of –**b** at terminal point of **a**. The sum is the vector joining the initial point of **a** to the final point of –**b**, which results in **c**. This figure represents **a** – **b** = **c**:

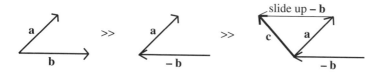

- The *sum of two vectors* $\overrightarrow{AB}$ and $\overrightarrow{CD}$ is written:

$$\overrightarrow{AB} + \overrightarrow{CD} = \overrightarrow{AD}$$

The sum of two vectors **A** and **B** is written:

$$\mathbf{A} + \mathbf{B} = \begin{bmatrix} a_1 \\ a_2 \end{bmatrix} + \begin{bmatrix} b_1 \\ b_2 \end{bmatrix} = \begin{bmatrix} a_1 + b_1 \\ a_2 + b_2 \end{bmatrix}$$

If $\mathbf{A} = \begin{bmatrix} 2 \\ 3 \end{bmatrix}$ and $\mathbf{B} = \begin{bmatrix} 3 \\ 4 \end{bmatrix}$ then $\mathbf{A} + \mathbf{B} = \begin{bmatrix} 5 \\ 7 \end{bmatrix}$

- Two *vectors can be added or subtracted* and expressed using unit vectors. If vector $\mathbf{A} = a_1\mathbf{i} + a_2\mathbf{j}$ and vector $\mathbf{B} = b_1\mathbf{i} + b_2\mathbf{j}$, then:

$$\mathbf{A} + \mathbf{B} = a_1\mathbf{i} + a_2\mathbf{j} + b_1\mathbf{i} + b_2\mathbf{j} = (a_1 + b_1)\mathbf{i} + (a_2 + b_2)\mathbf{j}$$
$$\mathbf{A} - \mathbf{B} = a_1\mathbf{i} + a_2\mathbf{j} - (b_1\mathbf{i} + b_2\mathbf{j}) = (a_1 - b_1)\mathbf{i} + (a_2 - b_2)\mathbf{j}$$

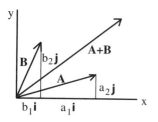

If **A** = 2**i** + 3**j** and **B** = 3**i** + 4**j**, then **A** + **B** = 5**i** + 7**j**.

• **Example:** Consider a ship moving along the ocean at a velocity **v** = 15 km/hr relative to the water, which has a current **c** = 2 km/hr. An angle θ = 45° exists between the direction of the ship and the direction of the ocean current.

The true velocity of the ship with respect to land is equal to the sum of the two vectors **v** + **c**.

To calculate the actual speed of the ship relative to land, set the velocity of the ship along the X-axis so that **v** = (15)**i** and the ocean current |**c**| = 2. Therefore:

   c = (2 cos 45°)**i** + (2 sin 45°)**j** ≈ 1.4**i** + 1.4**j**

The actual velocity **S** of the ship relative to land is:

   S = **v** + **c** = 15**i** + 1.4**i** + 1.4**j** = 16.4**i** + 1.4**j**

Therefore, the speed of the ship relative to land is:

   $|\mathbf{S}| = \sqrt{(16.4)^2 + (1.4)^2} = 16.46$ km/hr.

The angle the ship is deviating from **v** along the X-axis due to the current is: θ = tan⁻¹(1.4/16.4) = 4.9° = 0.0085 radians.

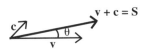

5.2 Introduction to Matrices

• This section includes definitions, notation, types of matrices including square, transpose, symmetric, and skew, and addition of matrices.

- A *matrix* is a rectangular *array* of numbers or functions. If a matrix has a single row or column, it is a *vector*. The components in a vector are called *elements* in a matrix. The array defining a matrix is enclosed in brackets and each number or function is called an *element* or *entry*.

- If a matrix has an equal number of rows and columns, it is a *square matrix*:

$$\begin{bmatrix} a & b \\ c & d \end{bmatrix}$$

- Matrices are often used to represent and *solve a set of equations*. The coefficients of the equations are the elements of a *coefficient matrix*. For the set of equations:

$$2x - 3y + z = 0$$
$$3x + y + 2z = 0$$
$$x + 2y + 2z = 0$$

The coefficient matrix is:

$$\begin{bmatrix} 2 & -3 & 1 \\ 3 & 1 & 2 \\ 1 & 2 & 2 \end{bmatrix}$$

- *Notation for matrices* includes boldface capital letters, writing an array of numbers or functions in brackets, or using double-subscript notation as shown below. When matrices and vectors are used together, the vectors are often represented using lower-case boldface letters and the matrices in upper-case boldface letters. In a matrix the rows and columns are often denoted as m number of rows and n number of columns resulting in an m by n matrix. An m by n matrix **A** is can be represented as:

$$\mathbf{A} = [a_{jk}] = \begin{bmatrix} a_{11} & a_{12} & \cdots & a_{1n} \\ a_{21} & a_{22} & \cdots & a_{2n} \\ \cdot & \cdot & \cdot & \cdot \\ a_{m1} & a_{m2} & \cdots & a_{mn} \end{bmatrix}$$

Using the a_{jk} *double-subscript notation*, the first subscript represents the row and the second subscript represents the column. For example, a_{32} represents the element located in the third row and the second column. If matrix **A** has m = n, then it is a square matrix. The elements in the *main diagonal* of a matrix are represented by: a_{11}, a_{22}, a_{33}, ... a_{mn}. A *submatrix* of matrix **A** has rows and/or columns absent.

• When a matrix is *transposed* (indicated by "T"), its rows and columns are changed so that the first row becomes the first column and the second row becomes the second column, and so on. The *transpose of matrix* **A**:

$$\mathbf{A} = [a_{jk}] = \begin{bmatrix} a_{11} & a_{12} & \cdots & a_{1n} \\ a_{21} & a_{22} & \cdots & a_{2n} \\ \cdot & \cdot & \cdot & \cdot \\ a_{m1} & a_{m2} & \cdots & a_{mn} \end{bmatrix}$$

is:

$$\mathbf{A}^T = [a_{kj}] = \begin{bmatrix} a_{11} & a_{21} & \cdots & a_{m1} \\ a_{12} & a_{22} & \cdots & a_{m2} \\ \cdot & \cdot & \cdot & \cdot \\ a_{1n} & a_{2n} & \cdots & a_{mn} \end{bmatrix}$$

• For example, if $\mathbf{A} = \begin{bmatrix} 1 & 2 \\ 3 & 4 \end{bmatrix}$, then $\mathbf{A}^T = \begin{bmatrix} 1 & 3 \\ 2 & 4 \end{bmatrix}$.

• If $\mathbf{A}^T = \mathbf{A}$, then **A** is called a *symmetric matrix*. If $\mathbf{A}^T = -\mathbf{A}$, then **A** is called a *skew-symmetric matrix*.

• *Properties of transpose matrices* include:

$(\mathbf{A} + \mathbf{B})^T = \mathbf{A}^T + \mathbf{B}^T$

$(c\mathbf{A})^T = c\mathbf{A}^T$, where c is a scalar.

• Two matrices $\mathbf{A} = [a_{kj}]$ and $\mathbf{B} = [b_{kj}]$ are equal if they have the same size and the corresponding elements are equal. Therefore in *equal matrices*, $a_{11} = b_{11}$, $a_{22} = b_{22}$, $a_{33} = b_{33}$, etc.

- *Matrices can be added or subtracted* if they are the same size by adding or subtracting the corresponding elements. For example, add **C** + **D**.

$$\mathbf{C} + \mathbf{D} = \begin{bmatrix} 1 & 2 \\ 3 & 4 \end{bmatrix} + \begin{bmatrix} 2 & -1 \\ 3 & 5 \end{bmatrix} = \begin{bmatrix} 3 & 1 \\ 6 & 9 \end{bmatrix}$$

- *Properties of addition of vectors and matrices* where **A** and **B** are the same size include:

$(\mathbf{A} + \mathbf{B}) + \mathbf{C} = \mathbf{A} + (\mathbf{B} + \mathbf{C})$ (associative)

$\mathbf{A} + \mathbf{B} = \mathbf{B} + \mathbf{A}$ (commutative)

$\mathbf{A} + \mathbf{A} = 2\mathbf{A}$

$\mathbf{A} + \mathbf{0} = \mathbf{A}$

$\mathbf{A} + (-\mathbf{A}) = \mathbf{0}$

- Note: The following properties are true for *differentiating*:

$$\frac{d}{dt}(f\mathbf{A}) = f\frac{d\mathbf{A}}{dt} + \mathbf{A}\frac{df}{dt}$$

$$\frac{d}{dt}(\mathbf{A} + \mathbf{B}) = \frac{d\mathbf{A}}{dt} + \frac{d\mathbf{B}}{dt}$$

5.3 Multiplication of Vectors and Matrices

- This section includes multiplication of vectors and matrices with scalars, multiplication of two matrices, multiplication of a vector with a matrix, and multiplication of row and column vectors.

- The following are *properties of multiplication of matrices* **A**, **B**, **C** and scalar c:

$(\mathbf{A} + \mathbf{B})\mathbf{C} = \mathbf{AC} + \mathbf{BC}$

$\mathbf{C}(\mathbf{A} + \mathbf{B}) = \mathbf{CA} + \mathbf{CB}$

$(c\mathbf{A})\mathbf{B} = c\mathbf{AB} = \mathbf{A}(c\mathbf{B})$

$\mathbf{A}(\mathbf{BC}) = (\mathbf{AB})\mathbf{C}$

$(c_1 + c_2)\mathbf{A} = c_1\mathbf{A} + c_2\mathbf{A}$

$c(\mathbf{A} + \mathbf{B}) = c\mathbf{A} + c\mathbf{B}$

$c_1(c_2)\mathbf{A} = (c_1c_2)\mathbf{A}$

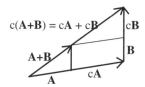

• *Multiplying vector A with scalar c* results in a vector having a magnitude of $|c||A|$ and a direction of **A**, where $|c|$ represents the absolute value of the scalar c and $|A|$ represents the magnitude of vector **A**. When c > 0, the displacement vector c**A** is parallel to **A** and pointing in the same direction as **A**. When c < 0, the vector c**A** is parallel to **A** but pointing in the opposite direction as **A**.

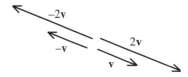

• When *multiplying matrices and scalars*, each element in the matrix is multiplied with the scalar. If matrix **A** = $[a_{jk}]$ is multiplied with scalar c, where $_j$ goes from 1 to 4 and $_k$ goes from 1 to 4, it can be written:

$$cA = [ca_{jk}] = \begin{bmatrix} ca_{11} & ca_{12} & ca_{13} & ca_{14} \\ ca_{21} & ca_{22} & ca_{23} & ca_{24} \\ ca_{31} & ca_{32} & ca_{33} & ca_{34} \\ ca_{41} & ca_{42} & ca_{43} & ca_{44} \end{bmatrix}$$

• To *multiply two matrices*, the number of columns in the first matrix must be equal to the number of rows in the second matrix. For matrix **A** = $[a_{jk}]$, which is an m by n matrix having m rows and n columns and matrix **B** = $[b_{jk}]$, which is a p by q matrix having p rows and q columns, the product exists if n = p.

The process for *multiplying two matrices* **A** and **B**, is to multiply each element in the first row with the corresponding element in the first column, then multiply each element in the second row with the corresponding element in the second column, and so on. Thereby multiplying each element in the jth row with the corresponding element in the kth column. If **C** is the product of matrices **A** and **B**, it can be written:

$$c_{jk} = \sum_{l=1}^{n} a_{jl}b_{lk} = a_{j1}b_{1k} + a_{j2}b_{2k} + \ldots + a_{jn}b_{nk}$$

where $j = 1, \ldots, m$ and $k = 1, \ldots, q$.

C can also be given by **AB = C** in matrix arrays:

$$
\begin{bmatrix}
a_{11} & a_{12} & a_{13} & a_{14} \\
a_{21} & a_{22} & a_{23} & a_{24} \\
a_{31} & a_{32} & a_{33} & a_{34} \\
a_{41} & a_{42} & a_{43} & a_{44}
\end{bmatrix}
\begin{bmatrix}
b_{11} & b_{12} & b_{13} & b_{14} \\
b_{21} & b_{22} & b_{23} & b_{24} \\
b_{31} & b_{32} & b_{33} & b_{34} \\
b_{41} & b_{42} & b_{43} & b_{44}
\end{bmatrix}
=
$$

m rows, n columns   p rows, q columns

$$
=
\begin{bmatrix}
c_{11} & c_{12} & c_{13} & c_{14} \\
c_{21} & c_{22} & c_{23} & c_{24} \\
c_{31} & c_{32} & c_{33} & c_{34} \\
c_{41} & c_{42} & c_{43} & c_{44}
\end{bmatrix}
$$

m rows, q columns

The highlighted row and column depicts the order that multiplication is carried out.

- The following represents the process of *multiplying two matrices*:

$$
\begin{bmatrix} \text{row } 1 \\ \text{row } 2 \end{bmatrix}
\begin{bmatrix} \text{column } 1 & \text{column } 2 \end{bmatrix}
$$

$$
=
\begin{bmatrix}
\text{row } 1 \bullet \text{column } 1 & \text{row } 1 \bullet \text{column } 2 \\
\text{row } 2 \bullet \text{column } 1 & \text{row } 2 \bullet \text{column } 2
\end{bmatrix}
$$

Note that the result is a matrix of *dot products* (see Section 5.4 for the dot product).

For example, multiply the following two matrices:

$$
\begin{bmatrix} 8 & 3 \\ 2 & 0 \end{bmatrix}
\begin{bmatrix} 1 & 4 \\ 2 & 5 \end{bmatrix}
=
\begin{bmatrix} 8 \cdot 1 + 3 \cdot 2 & 8 \cdot 4 + 3 \cdot 5 \\ 2 \cdot 1 + 0 \cdot 2 & 2 \cdot 4 + 0 \cdot 5 \end{bmatrix}
=
\begin{bmatrix} 14 & 47 \\ 2 & 8 \end{bmatrix}
$$

• In general **AB** ≠ **BA**, however they can be equal. Also, **AB** can equal zero even if neither **A** nor **B** is zero.

• When *multiplying a matrix and a vector*, the rule for two matrices applies, where the number of columns in the first matrix must be equal to the number of rows in the second matrix. Therefore, the number of columns in the matrix must be equal to the number of elements in the vector. For example:

$$\begin{bmatrix} 8 & 3 \\ 2 & 0 \end{bmatrix}\begin{bmatrix} 1 \\ 2 \end{bmatrix} = \begin{bmatrix} 8\cdot1+3\cdot2 \\ 2\cdot1+0\cdot2 \end{bmatrix} = \begin{bmatrix} 14 \\ 2 \end{bmatrix}$$

Note that $\begin{bmatrix} 1 \\ 2 \end{bmatrix}\begin{bmatrix} 8 & 3 \\ 2 & 0 \end{bmatrix}$ is undefined.

• Following are some examples of *multiplying row and column vectors*:

$$[3 \quad 4]\begin{bmatrix} 1 \\ 2 \end{bmatrix} = [3 + 8] = [11]$$

$$\begin{bmatrix} 1 \\ 2 \end{bmatrix}[3 \quad 4] = \begin{bmatrix} 3 & 4 \\ 6 & 8 \end{bmatrix}$$

5.4. Dot or Scalar Products

• This section includes equations that define the dot product, the dot product of parallel and perpendicular vectors, and properties of the dot product.

• The *dot product* (also called the *scalar product* or *inner product*) of two vectors is defined as:

A • **B** = |**A**||**B**| cos θ

where |**A**| and |**B**| represent the *magnitudes* of vectors **A** and **B** and θ is the angle between vectors **A** and **B**.

- Vectors **A** and **B** are *perpendicular* if **A** • **B** = 0, providing **A** or **B** does not equal zero. This is true because cos 90° = cos(π/2) = 0. For example, the dot product of **i** = (1, 0, 0) and **j** = (0, 1, 0) is **i** • **j** = 0, because **i** and **j** are perpendicular to each other.

- Vectors **A** and **B** are *parallel* if **A** • **B** = |**A**| |**B**|, providing **A** or **B** does not equal zero. This is true because cos 0 = 1. The dot product of **i** = (1, 0, 0) and **i** = (1, 0, 0) is **i** • **i** = 1.

- The *dot product of a vector with itself* **A** • **A**, has θ = 0 and because cos 0 = 1, then **A** • **A** = |**A**|^2 = length-squared.

- The dot product can also be used to compute cos θ:

$$\cos\theta = \frac{\mathbf{A} \cdot \mathbf{B}}{|\mathbf{A}||\mathbf{B}|}$$

- The dot product written in the form of |**A**||**B**| cos θ represents **A** • **B** without coordinates. The dot product can also be written in the form [$a_1b_1 + a_2b_2$] that does involve coordinates.

- The dot or scalar product of vector **A** = a_1**i** + a_2**j** and vector **B** = b_1**i** + b_2**j** can be written as:

$$\mathbf{A} \cdot \mathbf{B} = \begin{bmatrix} a_1 \\ a_2 \end{bmatrix} \cdot \begin{bmatrix} b_1 \\ b_2 \end{bmatrix} = a_1b_1 + a_2b_2$$

Or equivalently:

$$\mathbf{A} \cdot \mathbf{B} = (a_1\mathbf{i} + a_2\mathbf{j}) \cdot (b_1\mathbf{i} + b_2\mathbf{j})$$
$$= a_1b_1\mathbf{i} \cdot \mathbf{i} + a_1b_2\mathbf{i} \cdot \mathbf{j} + a_2b_1\mathbf{j} \cdot \mathbf{i} + a_2b_2\mathbf{j} \cdot \mathbf{j}$$
$$= a_1b_1(1) + a_1b_2(0) + a_2b_1(0) + a_2b_2(1)$$
$$= a_1b_1 + a_2b_2$$

The terms with **i** • **j** and **j** • **i** equal zero and **i** • **i** and **j** • **j** equal one. Therefore:

$$\mathbf{A} \cdot \mathbf{B} = (a_1\mathbf{i} + a_2\mathbf{j}) \cdot (b_1\mathbf{i} + b_2\mathbf{j}) = a_1b_1 + a_2b_2$$

In three-dimensions the dot product of **A** and **B** is:

$$\mathbf{A} \cdot \mathbf{B} = a_1b_1 + a_2b_2 + a_3b_3$$

- In summary, *unit vectors combine* as follows:

  $\mathbf{i} \cdot \mathbf{i} = \mathbf{j} \cdot \mathbf{j} = \mathbf{k} \cdot \mathbf{k} = 1$

  $\mathbf{i} \cdot \mathbf{j} = \mathbf{i} \cdot \mathbf{k} = \mathbf{j} \cdot \mathbf{i} = \mathbf{j} \cdot \mathbf{k} = \mathbf{k} \cdot \mathbf{i} = \mathbf{k} \cdot \mathbf{j} = 0$

- One application of the dot product is to find the *angle between two vectors*. If vector $\mathbf{A} = \langle 2, 3 \rangle = (2\mathbf{i} + 3\mathbf{j})$ and $\mathbf{B} = \langle 3, 4 \rangle = (3\mathbf{i} + 4\mathbf{j})$:

The angle θ between vectors $\mathbf{A}$ and $\mathbf{B}$ can be found using:

$$\cos \theta = \frac{\mathbf{A} \cdot \mathbf{B}}{|\mathbf{A}||\mathbf{B}|} = \frac{\langle 2, 3 \rangle \cdot \langle 3, 4 \rangle}{\sqrt{2^2 + 3^2}\sqrt{3^2 + 4^2}} = \frac{(2)(3) + (3)(4)}{\sqrt{13}\sqrt{25}} = \frac{18}{5\sqrt{13}}$$

Therefore, $\theta = \arccos \dfrac{18}{5\sqrt{13}} \approx 3.2°$

- Another application of the dot product is the dot product of *force* $\mathbf{F}$ and *distance* $\mathbf{d}$, which equals *work* done W: $\mathbf{F} \cdot \mathbf{d} = W$ where $\mathbf{F}$ is acting on an object to displace it. Therefore, work done can be written:

  $W = \mathbf{F} \cdot \mathbf{d} = |\mathbf{F}||\mathbf{d}| \cos \theta$

where $\mathbf{F}$ is acting on an object to displace it by distance $\mathbf{d}$.

The work done by $\mathbf{F}$ in displacement is the *magnitude* $|\mathbf{F}|$ of the force multiplied by length $|\mathbf{d}|$ of the displacement multiplied with cosine of the angle θ between $\mathbf{F}$ and $\mathbf{d}$. The work is zero if $\mathbf{F}$ and $\mathbf{d}$ are perpendicular to each other.

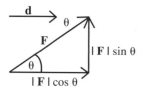

If $\theta = 45°$ then,

$$W = \mathbf{F} \cdot \mathbf{d} = |\mathbf{F}||\mathbf{d}| \cos 45° = |\mathbf{F}||\mathbf{d}| \sqrt{2}/2$$

If $\theta = 90°$ then,

$$W = \mathbf{F} \cdot \mathbf{d} = |\mathbf{F}||\mathbf{d}| \cos 90° = |\mathbf{F}||\mathbf{d}|(0) = 0$$

• Some *properties of the dot product* of vectors **A**, **B**, **C** and scalar c are:

$$c(\mathbf{A} \cdot \mathbf{B}) = (c\mathbf{A}) \cdot \mathbf{B} = \mathbf{A} \cdot (c\mathbf{B})$$
$$\mathbf{A} \cdot (\mathbf{B} + \mathbf{C}) = (\mathbf{A} \cdot \mathbf{B}) + (\mathbf{A} \cdot \mathbf{C})$$
$$\mathbf{A} \cdot \mathbf{B} = \mathbf{B} \cdot \mathbf{A}$$

• The following property applies for *differentiating*:

$$\frac{d}{dt}\mathbf{A} \cdot \mathbf{B} = \mathbf{A} \cdot \frac{d\mathbf{B}}{dt} + \frac{d\mathbf{A}}{dt} \cdot \mathbf{B}$$

5.5 Vector or Cross Product

• This section includes equations that define the cross product, the cross product of two vectors, minimum and maximum values, and applications of the cross product.

• The *vector product or cross product* of two vectors is defined as:

$$\mathbf{A} \times \mathbf{B} = |\mathbf{A}||\mathbf{B}| \sin \theta$$

where $|\mathbf{A}|$ and $|\mathbf{B}|$ represent the *magnitudes* (or *lengths*) of vectors **A** and **B** and θ is the angle between vectors **A** and **B**. The product exists in three dimensions with **A** and **B** in a plane and $\mathbf{A} \times \mathbf{B}$ normal to the plane. The *cross product* of two vectors produces a third vector with a length of $|\mathbf{A}||\mathbf{B}| \sin \theta$ and a direction perpendicular to **A** and **B**. The length of $\mathbf{A} \times \mathbf{B}$ depends on $\sin \theta$ and is greatest when $\theta = 90°$ or $\sin \theta = 1$.

• The cross product of two vectors occurs geometrically according to what is referred to as the *right-hand screw rule*. This rule denotes that when taking the cross product $\mathbf{A} \times \mathbf{B}$ and moving from vector **A** to vector **B** through angle θ results in vector $\mathbf{A} \times \mathbf{B}$, which is perpendicular to both

A and **B**. The right-hand rule can be visualized by curling the fingers of the right hand from **A** to **B**, where **A** × **B** points in the direction of the right thumb. Conversely, for the cross product **B** × **A**, moving from vector **B** to vector **A** through angle θ results in a vector perpendicular to both **A** and **B** but pointing in the opposite direction of **A** × **B**. Therefore, by the right-hand rule, **A** × **B** and **B** × **A** point in opposite directions but have the same magnitude.

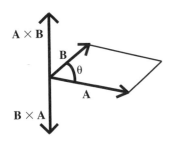

- The *vector or cross product* of two vectors written in terms of their components as vector **A** = $a_1\mathbf{i} + a_2\mathbf{j} + a_3\mathbf{k}$ and vector **B** = $b_1\mathbf{i} + b_2\mathbf{j} + b_3\mathbf{k}$ can be calculated using the same procedure as when calculating a *determinant*. (See Section 5.6. for a summary of determinants.)

$$\mathbf{A} \times \mathbf{B} = (a_1\mathbf{i} + a_2\mathbf{j} + a_3\mathbf{k}) \times (b_1\mathbf{i} + b_2\mathbf{j} + b_3\mathbf{k}) = \begin{vmatrix} \mathbf{i} & \mathbf{j} & \mathbf{k} \\ a_1 & a_2 & a_3 \\ b_1 & b_2 & b_3 \end{vmatrix}$$

$$= (a_2b_3 - a_3b_2)\mathbf{i} + (a_3b_1 - a_1b_3)\mathbf{j} + (a_1b_2 - a_2b_1)\mathbf{k}$$
$$= (a_2b_3 - a_3b_2)\mathbf{i} - (a_1b_3 - a_3b_1)\mathbf{j} + (a_1b_2 - a_2b_1)\mathbf{k}$$

The terms with **i** × **i**, **j** × **j**, and **k** × **k** equal zero. The **i**, **j**, **k** *unit vectors combine* as follows:

$$\mathbf{i} \times \mathbf{i} = \mathbf{j} \times \mathbf{j} = \mathbf{k} \times \mathbf{k} = 0$$
$$\mathbf{i} \times \mathbf{j} = \mathbf{k}, \mathbf{i} \times \mathbf{k} = -\mathbf{j}$$
$$\mathbf{j} \times \mathbf{i} = -\mathbf{k}, \mathbf{j} \times \mathbf{k} = \mathbf{i}$$
$$\mathbf{k} \times \mathbf{i} = \mathbf{j}, \mathbf{k} \times \mathbf{j} = -\mathbf{i}$$

Considering the nature of how the unit vectors combine, $\mathbf{A} \times \mathbf{B}$ can be written out as:

$$(a_1\mathbf{i} + a_2\mathbf{j} + a_3\mathbf{k}) \times (b_1\mathbf{i} + b_2\mathbf{j} + b_3\mathbf{k})$$
$$= a_1b_1\mathbf{i} \times \mathbf{i} + a_1b_2\mathbf{i} \times \mathbf{j} + a_1b_3\mathbf{i} \times \mathbf{k}$$
$$+ a_2b_1\mathbf{j} \times \mathbf{i} + a_2b_2\mathbf{j} \times \mathbf{j} + a_2b_3\mathbf{j} \times \mathbf{k}$$
$$+ a_3b_1\mathbf{k} \times \mathbf{i} + a_3b_2\mathbf{k} \times \mathbf{j} + a_3b_3\mathbf{k} \times \mathbf{k}$$
$$= \mathbf{0} + a_1b_2\mathbf{k} + a_1b_3(-\mathbf{j}) + a_2b_1(-\mathbf{k}) + \mathbf{0} + a_2b_3\mathbf{i} + a_3b_1\mathbf{j} + a_3b_2(-\mathbf{i}) + \mathbf{0}$$
$$= (a_2b_3 - a_3b_2)\mathbf{i} + (a_3b_1 - a_1b_3)\mathbf{j} + (a_1b_2 - a_2b_1)\mathbf{k}$$

• The unit vectors $\mathbf{i}$, $\mathbf{j}$, and $\mathbf{k}$ are *perpendicular* to each other. Therefore, the angle between $\mathbf{i}$ and $\mathbf{j}$ is $\pi/2$ and by the right-hand rule the cross product of $\mathbf{i}$ and $\mathbf{j}$ is:

$$\mathbf{i} \times \mathbf{j} = |\mathbf{i}|\,|\mathbf{j}|\sin(\pi/2) = \mathbf{k}$$

The cross product of $\mathbf{i}$ with itself is:

$$\mathbf{i} \times \mathbf{i} = |\mathbf{i}|\,|\mathbf{i}|\sin 0 = \mathbf{0}.$$

• The *maximum value of the cross product* of two vectors occurs when the angle θ is $\pi/2$ and $\sin \pi/2 = 1$. Therefore the two *vectors are perpendicular* to each other. Conversely, the *minimum value of the cross product* of two vectors occurs when the angle θ is 0 or π and $\sin 0 = \sin \pi = 0$, and, therefore the two *vectors are parallel*.

• The cross product can represent the area of a parallelogram with sides $\mathbf{A}$ and $\mathbf{B}$, where the value resulting from $\mathbf{A} \times \mathbf{B}$ is both the *length of vector $\mathbf{A} \times \mathbf{B}$* and the *area of the parallelogram*. The length of the cross product is the area and the area of the parallelogram is $|\mathbf{A} \times \mathbf{B}|$, which is the magnitude of the area. A parallelogram with sides $\mathbf{A}$ and $\mathbf{B}$, has area $|a_1b_2 - a_2b_1|$. In an XY plane $\mathbf{A} \times \mathbf{B} = (a_1b_2 - a_2b_1)\mathbf{k}$.

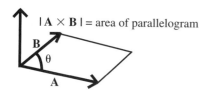

$|\mathbf{A} \times \mathbf{B}| = $ area of parallelogram

• One important application of the *cross product* is *torque*, which is a force acting on an object to cause rotation. A force **F** can be applied to a lever arm or a radius vector **r**, which has its initial point located at the origin of rotation and causes the object to rotate. The torque is a vector having a magnitude that measures the rotation of the force and has a direction of the axis of rotation. The cross product **F** $\times$ **r** = **T** is the torque of the force about the origin for a force **F** acting at a point with position vector **r**.

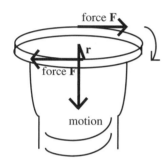

• An application of the *cross product* and *dot product* together is the *volume of a parallelepiped*. A parallelepiped with sides given by vectors **A**, **B**, and **C** is represented by:

$$|\mathbf{A} \cdot \mathbf{B} \times \mathbf{C}| =$$
$$(a_1\mathbf{i} + a_2\mathbf{j} + a_3\mathbf{k}) \cdot (b_1\mathbf{i} + b_2\mathbf{j} + b_3\mathbf{k}) \times (c_1\mathbf{i} + c_2\mathbf{j} + c_3\mathbf{k}) =$$

$$\begin{vmatrix} a_1 & a_2 & a_3 \\ b_1 & b_2 & b_3 \\ c_1 & c_2 & c_3 \end{vmatrix} = a_1(b_2c_3 - b_3c_2) + a_2(b_3c_1 - b_1c_3) + a_3(b_1c_2 - b_2c_1)$$

$$= |\mathbf{B} \times \mathbf{C}| \, |\mathbf{A}| \cos \theta$$

where $|\mathbf{B} \times \mathbf{C}|$ is area of base and $|\mathbf{A}| \cos \theta$ is height.

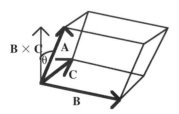

- The *volume of a cube* can be represented using the unit vectors **i, j, k**:

$$\begin{vmatrix} 1 & 0 & 0 \\ 0 & 1 & 0 \\ 0 & 0 & 1 \end{vmatrix} = 1$$

- *Properties of the cross product* involving vectors **A, B, C** and scalar c include:

$$\mathbf{A} \times \mathbf{B} = -(\mathbf{B} \times \mathbf{A})$$
$$c(\mathbf{A} \times \mathbf{B}) = (c\mathbf{A}) \times \mathbf{B} = \mathbf{A} \times (c\mathbf{B})$$
$$\mathbf{A} \times (\mathbf{B} + \mathbf{C}) = (\mathbf{A} \times \mathbf{B}) + (\mathbf{A} \times \mathbf{C})$$
$$\mathbf{A} \cdot \mathbf{B} \times \mathbf{C} = \mathbf{A} \times \mathbf{B} \cdot \mathbf{C} = \mathbf{B} \cdot \mathbf{C} \times \mathbf{A} = \mathbf{C} \cdot \mathbf{A} \times \mathbf{B}$$
$$|\mathbf{A} \cdot \mathbf{B}|^2 + |\mathbf{A} \times \mathbf{B}|^2 = |\mathbf{A}|^2 |\mathbf{B}|^2 \cos^2\theta + |\mathbf{A}|^2 |\mathbf{B}|^2 \sin^2\theta$$
$$= |\mathbf{A}|^2 |\mathbf{B}|^2$$

- The following property applies to *differentiating*:

$$\frac{d}{dt} \mathbf{A} \times \mathbf{B} = \mathbf{A} \times \frac{d\mathbf{B}}{dt} + \frac{d\mathbf{A}}{dt} \times \mathbf{B}$$

5.6 Summary of Determinants

- This section provides a brief review of determinants including definitions and using the method of determinants and Cramer's rule to solve systems of equations.

- A two-by-two *determinant* is written as follows:

$$\begin{vmatrix} a_1 & a_2 \\ b_1 & b_2 \end{vmatrix} = a_1 b_2 - a_2 b_1$$

- A three-by-three determinant is written as follows:

$$\begin{vmatrix} a_1 & a_2 & a_3 \\ b_1 & b_2 & b_3 \\ c_1 & c_2 & c_3 \end{vmatrix} = a_1 \begin{vmatrix} b_2 & b_3 \\ c_2 & c_3 \end{vmatrix} - a_2 \begin{vmatrix} b_1 & b_3 \\ c_1 & c_3 \end{vmatrix} + a_3 \begin{vmatrix} b_1 & b_2 \\ c_1 & c_2 \end{vmatrix}$$
$$= a_1(b_2 c_3 - b_3 c_2) - a_2(b_1 c_3 - b_3 c_1) + a_3(b_1 c_2 - b_2 c_1)$$

• *Two equations* with two unknown variables can be solved using the *method of determinants* and *Cramer's Rule*. The two equations can be represented by:

$$a_1x + b_1y = c_1$$
$$a_2x + b_2y = c_2$$

where a, b, and c represent known coefficients or constants and x and y are unknown variables.

First create three *matrices of coefficients* D, D_x, and D_y and calculate the determinants:

$$D = \begin{vmatrix} a_1 & b_1 \\ a_2 & b_2 \end{vmatrix} = a_1b_2 - a_2b_1$$

$$D_x = \begin{vmatrix} c_1 & b_1 \\ c_2 & b_2 \end{vmatrix} = c_1b_2 - c_2b_1$$

$$D_y = \begin{vmatrix} a_1 & c_1 \\ a_2 & c_2 \end{vmatrix} = a_1c_2 - a_2c_1$$

The solutions for x and y are:

$$x = D_x/D \text{ and } y = D_y/D, \text{ providing } D \neq 0.$$

• *Three equations* with three unknown variables can also be solved using the *method of determinants* and *Cramer's Rule*. The three equations can be represented by:

$$a_1x + b_1y + c_1z = d_1$$
$$a_2x + b_2y + c_2z = d_2$$
$$a_3x + b_3y + c_3z = d_3$$

where a, b, c, and d represent known coefficients or constants and x, y, and z are unknown variables.

First create four matrices of coefficients D, D_x, D_y, and D_z and calculate the determinants:

$$D = \begin{vmatrix} a_1 & b_1 & c_1 \\ a_2 & b_2 & c_2 \\ a_3 & b_3 & c_3 \end{vmatrix}$$

$$= a_1(b_2c_3 - b_3c_2) - a_2(b_1c_3 - b_3c_1) + a_3(b_1c_2 - b_2c_1)$$

$$D_x = \begin{vmatrix} d_1 & b_1 & c_1 \\ d_2 & b_2 & c_2 \\ d_3 & b_3 & c_3 \end{vmatrix}$$

$$= d_1(b_2c_3 - b_3c_2) - d_2(b_1c_3 - b_3c_1) + d_3(b_1c_2 - b_2c_1)$$

$$D_y = \begin{vmatrix} a_1 & d_1 & c_1 \\ a_2 & d_2 & c_2 \\ a_3 & d_3 & c_3 \end{vmatrix}$$

$$= a_1(d_2c_3 - d_3c_2) - a_2(d_1c_3 - d_3c_1) + a_3(d_1c_2 - d_2c_1)$$

$$D_z = \begin{vmatrix} a_1 & b_1 & d_1 \\ a_2 & b_2 & d_2 \\ a_3 & b_3 & d_3 \end{vmatrix}$$

$$= a_1(b_2d_3 - b_3d_2) - a_2(b_1d_3 - b_3d_1) + a_3(b_1d_2 - b_2d_1)$$

The solutions for x, y, and z are:

$x = D_x/D$, $y = D_y/D$, and $z = D_z/D$, providing $D \neq 0$.

• Note that the values of the determinant are not affected if the determinant is *transposed*.

5.7 Matrices and Linear Algebra

• This section includes information about representing and solving systems of linear equations.

• Systems of *linear equations* can be solved using *matrices*. Solutions for two equations with two unknown variables exist where two lines intersect. Similarly, solutions for three equations with three unknown variables exist where three planes intersect. To solve a *system of n linear equations* with n unknown variables, matrices can be formed and the *method of determinants* as described in the previous paragraphs can be employed.

Other methods can be employed to solve systems of equations that go beyond the scope of this book. (See *Master Math: Algebra* Chapter 8 for a discussion of solving simple systems of two and three linear equations using various methods. Also, mathematics books dedicated to solving both linear and non-linear systems of equations should be consulted for a comprehensive discussion.)

• A *system of m linear equations with n unknowns* can be represented by $\mathbf{A}\mathbf{x} = \mathbf{d}$, where $\mathbf{A} = [a_{jk}]$ is the *coefficient matrix* containing given coefficients (or constants), $\mathbf{x} = x_1,...x_n$ is the *solution set* and $\mathbf{d} = d_1,...,d_n$ are given numbers. If d_i are all zero, the system of equations is called a *homogeneous system*. If at least one d_i is not zero, the system of equations is called a *non-homogeneous system*.

The system $\mathbf{A}\mathbf{x} = \mathbf{d}$ can be represented as:

$$\mathbf{A}\mathbf{x} = \mathbf{d} = \begin{bmatrix} a_{11} & a_{12} & \cdots & a_{1n} \\ a_{21} & a_{22} & \cdots & a_{2n} \\ \cdot & \cdot & \cdot & \cdot \\ a_{m1} & a_{m2} & \cdots & a_{mn} \end{bmatrix} \begin{bmatrix} x_1 \\ \cdot \\ \cdot \\ x_n \end{bmatrix} = \begin{bmatrix} d_1 \\ \cdot \\ \cdot \\ d_n \end{bmatrix}$$

For two equations and two unknowns:

$$a_{11}x_1 + a_{12}x_2 = d_1$$
$$a_{21}x_1 + a_{22}x_2 = d_2$$

$\mathbf{A}\mathbf{x}$ can be written:

$$\begin{bmatrix} a_{11} & a_{12} \\ a_{21} & a_{22} \end{bmatrix} \begin{bmatrix} x_1 \\ x_2 \end{bmatrix} = \begin{bmatrix} a_{11}x_1 + a_{12}x_2 \\ a_{21}x_1 + a_{22}x_2 \end{bmatrix} = x_1 \begin{bmatrix} a_{11} \\ a_{21} \end{bmatrix} + x_2 \begin{bmatrix} a_{12} \\ a_{22} \end{bmatrix}$$

The solution set for $\mathbf{A}\mathbf{x} = \mathbf{d}$ is given by $\mathbf{x} = \mathbf{A}^{-1}\mathbf{d}$.

- To find the solution set $\mathbf{x} = \mathbf{A}^{-1}\mathbf{d}$, the *inverse of a matrix* $\mathbf{A}^{-1}$ can be expressed using the *determinant* of $\mathbf{A}$. For example for two equations:

$$a_1x_1 + b_1x_2 = d_1$$
$$a_2x_1 + b_2x_2 = d_2$$

The solution set $\mathbf{x} = \mathbf{A}^{-1}\mathbf{d}$ is:

$$\mathbf{x} = \mathbf{A}^{-1}\mathbf{d} = \frac{1}{D}\begin{bmatrix} b_2 & -b_1 \\ -a_2 & a_1 \end{bmatrix}\begin{bmatrix} d_1 \\ d_2 \end{bmatrix} = \frac{1}{D}\begin{bmatrix} b_2d_1 & -b_1d_2 \\ -a_2d_1 & a_1d_2 \end{bmatrix}$$

where D is the determinant of $\mathbf{A}$ and is given by:

$$D = \begin{vmatrix} a_1 & b_1 \\ a_2 & b_2 \end{vmatrix} = a_1b_2 - a_2b_1$$

Remember, a matrix can be *transposed* and the determinant is unaffected.

- Note that the inverse matrix $\mathbf{A}^{-1}$ multiplied by the original matrix $\mathbf{A}$ is the *identity matrix* $\mathbf{I}$. The identity matrix has 1's on the diagonal and 0's elsewhere and behaves as the number 1. A two-by-two identity matrix is:

$$\mathbf{I} = \begin{bmatrix} 1 & 0 \\ 0 & 1 \end{bmatrix}$$

- In three-by-three matrices the inverse also uses the determinant D. The determinate of a three-by-three matrix is:

Determinant of $\mathbf{A} = \mathbf{a} \cdot \mathbf{b} \times \mathbf{c}$
$$= (a_1\mathbf{i} + a_2\mathbf{j} + a_3\mathbf{k}) \cdot (b_1\mathbf{i} + b_2\mathbf{j} + b_3\mathbf{k}) \times (c_1\mathbf{i} + c_2\mathbf{j} + c_3\mathbf{k})$$
$$= a_1(b_2c_3 - b_3c_2) + a_2(b_3c_1 - b_1c_3) + a_3(b_1c_2 - b_2c_1)$$

which represents *volume* of a box-shaped object.

Then for matrix $\mathbf{A} = \begin{bmatrix} a_1 & b_1 & c_1 \\ a_2 & b_2 & c_2 \\ a_3 & b_3 & c_3 \end{bmatrix}$

where the *column vectors* **a**, **b**, **c** represent the edges of the box extending from the origin, the inverse of **A** is given by:

$$\mathbf{A}^{-1} = \frac{1}{D}\begin{bmatrix} \mathbf{b}\times\mathbf{c} \\ \mathbf{c}\times\mathbf{a} \\ \mathbf{a}\times\mathbf{b} \end{bmatrix} = \frac{1}{D}\begin{bmatrix} b_2c_3-b_3c_2 & b_3c_1-b_1c_3 & b_1c_2-b_2c_1 \\ c_2a_3-c_3a_2 & c_3a_1-c_1a_3 & c_1a_2-c_2a_1 \\ a_2b_3-a_3b_2 & a_3b_1-a_1b_3 & a_1b_2-a_2b_1 \end{bmatrix}$$

Note that the first row of $\mathbf{A}^{-1}$ does not use the first column of **A** except in the calculation of 1/D.

Therefore, the solution to $\mathbf{x} = \mathbf{A}^{-1}\mathbf{d}$ is:

$$\mathbf{x} = \frac{1}{D}\begin{bmatrix} \mathbf{b}\times\mathbf{c} \\ \mathbf{c}\times\mathbf{a} \\ \mathbf{a}\times\mathbf{b} \end{bmatrix}\begin{bmatrix} \\ \mathbf{d} \\ \\ \end{bmatrix} = \frac{1}{D}\begin{bmatrix} \mathbf{d}\bullet(\mathbf{b}\times\mathbf{c}) \\ \mathbf{d}\bullet(\mathbf{c}\times\mathbf{a}) \\ \mathbf{d}\bullet(\mathbf{a}\times\mathbf{b}) \end{bmatrix}$$

The solutions for x, y, and z (or x_1, x_2, x_3) in three equations are ratios of *determinants*. Using *Cramer's Rule*:

$$x = |\mathbf{d}\bullet(\mathbf{b}\times\mathbf{c})|/|\mathbf{a}\bullet(\mathbf{b}\times\mathbf{c})|$$
$$y = |\mathbf{d}\bullet(\mathbf{c}\times\mathbf{a})|/|\mathbf{a}\bullet(\mathbf{b}\times\mathbf{c})|$$
$$z = |\mathbf{d}\bullet(\mathbf{a}\times\mathbf{b})|/|\mathbf{a}\bullet(\mathbf{b}\times\mathbf{c})|$$

where $D = |\mathbf{a}\bullet(\mathbf{b}\times\mathbf{c})|$

• The *inverse of a matrix* can also be found using *Gauss-Jordan elimination*, which combines *Gauss elimination* with an identity matrix. See a text on linear algebra for a discussion of this method.

• Another standard method for *solving a system of linear equations* is called *Gauss elimination*. To use this *elimination* method, transform the equations into a matrix, perform operations on the matrix until an *upper triangular matrix* is formed, transform the *triangular matrix* back into equation form, and solve for the unknown variables using substitution. Operations used in forming the upper triangular matrix include multiplying a row by a non-zero constant, interchanging two rows, and adding a multiple of one row to another.

- The general procedure for *Gauss elimination* is:

(a.) Transform the equations into a matrix by writing the coefficients of the equations into a matrix format:

$$a_1x + b_1y + c_1z = d_1$$
$$a_2x + b_2y + c_2z = d_2$$
$$a_3x + b_3y + c_3z = d_3$$

Then convert to a coefficient matrix:

$$\begin{vmatrix} a_1 & b_1 & c_1 & d_1 \\ a_2 & b_2 & c_2 & d_2 \\ a_3 & b_3 & c_3 & d_3 \end{vmatrix}$$

(b.) Create an upper triangular matrix by adding multiples of the coefficient rows to each other until an upper triangular matrix is formed. This involves, multiplying a row through by a non-zero constant, interchanging two rows and adding the multiple of one row to another row to yield a zero coefficient in the lower left of the upper triangular matrix. In a three-by-three upper triangular matrix, a_2, a_3, and b_3 must be converted to zeros and a_1, b_2, and c_3 must be converted to ones.

$$\begin{vmatrix} 1 & b_1 & c_1 & d_1 \\ 0 & 1 & c_2 & d_2 \\ 0 & 0 & 1 & d_3 \end{vmatrix}$$

(c.) Once the upper triangular matrix is formed, transform the coefficient matrix back into the form of the equations. The known variable (from the third equation in the three-by-three matrix) can be substituted back into the reduced equation to solve for the next variable, and then both known variables can be substituted into an original equation to find the third variable.

(d.) The results can be checked by substituting the variables into the original equations.

- **Example:** Solve the three equations for x, y, and z:

$$x + y + z = -1$$
$$2x - y + z = 0$$
$$-x + y - z = -2$$

In matrix form:

$$\begin{vmatrix} 1 & 1 & 1 & -1 \\ 2 & -1 & 1 & 0 \\ -1 & 1 & -1 & -2 \end{vmatrix}$$

Multiply row 1 by -2, then add it to row 2 to make the a_2 position zero:

row 1×-2 is:

$$-2 \quad -2 \quad -2 \quad 2$$

Add this new row 1 to row 2 to get new row 2:

$$0 \quad -3 \quad -1 \quad 2$$

Add row 1 and row 3 to make a_3 zero resulting in new row 3:

$$0 \quad 2 \quad 0 \quad -3$$

The new matrix is:

$$\begin{vmatrix} 1 & 1 & 1 & -1 \\ 0 & -3 & -1 & 2 \\ 0 & 2 & 0 & -3 \end{vmatrix}$$

Switch the second and third rows:

$$\begin{vmatrix} 1 & 1 & 1 & -1 \\ 0 & 2 & 0 & -3 \\ 0 & -3 & -1 & 2 \end{vmatrix}$$

Add row 2 and row 3 to get new row 2 and make b_2 to be 1:

$$0 \quad -1 \quad -1 \quad -1$$

Multiply new row 2 by –1:

$$\begin{vmatrix} 1 & 1 & 1 & -1 \\ 0 & 1 & 1 & 1 \\ 0 & -3 & -1 & 2 \end{vmatrix}$$

Add three times row 2 to row 3 to make b_3 to be 0:

$$0 \quad 0 \quad 2 \quad 5$$

The new matrix becomes:

$$\begin{vmatrix} 1 & 1 & 1 & -1 \\ 0 & 1 & 1 & 1 \\ 0 & 0 & 2 & 5 \end{vmatrix}$$

Divide row 3 by 2 to make c_3 to be 1. The upper triangular matrix becomes:

$$\begin{vmatrix} 1 & 1 & 1 & -1 \\ 0 & 1 & 1 & 1 \\ 0 & 0 & 1 & 5/2 \end{vmatrix}$$

The three equations take the form:

$$1x + 1y + 1z = -1$$
$$1y + 1z = 1$$
$$1z = 5/2$$

Therefore, $z = 5/2$.

Substitute z into equation 2:

$$1y + 5/2 = 1$$

Therefore, $y = -3/2$.

Substitute y and z into equation 1:

$$1x + (-3/2) + (5/2) = -1$$
$$1x + 1 = -1$$

Therefore, $x = -2$.

Therefore, x = –2, y = –3/2, and z = 5/2.

Check the results by substituting x, y, and z into original equations 1 and 2:

$$x + y + z = -1$$
$$1(-2) + (-3/2) + (5/2) = -1$$
$$-2 + 1 = -1$$
$$-1 = -1$$

$$2x - y + z = 0$$
$$2(-2) - -3/2 + 5/2 = 0$$
$$-4 + 8/2 = 0$$
$$-4 + 4 = 0$$
$$0 = 0$$

5.8 The Position Vector, Parametric Equations, Curves, and Surfaces

• This section provides general information about the position vector and parameterization of a line, a plane, a cylinder, a cone, a sphere, a circle, and a curve.

• A vector with its initial point at the origin of a *rectangular coordinate system* is called a *position vector*. A *position vector* is defined according to the location of its *terminal point*. A *position vector* can be used to locate the *position of a moving object* and can be written:

$$\mathbf{R}(t) = x(t)\mathbf{i} + y(t)\mathbf{j} + z(t)\mathbf{k}$$

• The *position vector* given by $\mathbf{R} = x\mathbf{i} + y\mathbf{j} + z\mathbf{k}$ existing between points (x_0,y_0,z_0) and (x_1,y_1,z_1) can be written:

$$\mathbf{R} = x_0\mathbf{i} + y_0\mathbf{j} + z_0\mathbf{k} + t(x_1 - x_0)\mathbf{i} + t(y_1 - y_0)\mathbf{j} + t(z_1 - z_0)\mathbf{k}$$

where t is a scalar often representing time. From this equation results the *parametric equations for a line*:

$$x = x_0 + t(x_1 - x_0)$$
$$y = y_0 + t(y_1 - y_0)$$
$$z = z_0 + t(z_1 - z_0)$$

These equations can also be written in the form:

$$\frac{x - x_0}{x_1 - x_0} = \frac{y - y_0}{y_1 - y_0} = \frac{z - z_0}{z_1 - z_0}$$

- Each point along this line $\mathbf{R} = \mathbf{R_0} + t\mathbf{v}$ where $\mathbf{R_0} = x_0\mathbf{i} + y_0\mathbf{j} + z_0\mathbf{k}$ can be evaluated by adding multiples of $\mathbf{v}$ to $\mathbf{R_0}$:

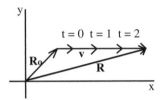

- A *point in a plane* having coordinates (x,y) (depicted below) can be represented using the *position vector* $\mathbf{R} = x\mathbf{i} + y\mathbf{j}$ with its terminal point on the point (x,y) and the *parametric equations* x = f(t), y = g(t). After substituting for x and y the vector equation becomes:

$$\mathbf{R} = \mathbf{F}(t) = f(t)\mathbf{i} + g(t)\mathbf{j}$$

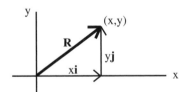

- To *parameterize a plane* that passes through a point $\mathbf{R_0} = (x_0, y_0, z_0)$ and has two non-parallel vectors $\mathbf{v_1}$ and $\mathbf{v_2}$ where all points on the plane can be identified by beginning at point P_0 and moving parallel by adding multiples of $\mathbf{v_1}$ and $\mathbf{v_2}$ to $\mathbf{R_0}$. A *plane* has two parameters. The parameters can be expressed as:

$$x = t_1, y = t_2, z = f(t_1, t_2)$$

where t_1 and t_2 represent the parameters.

The parametric equation can be represented by:

$$\mathbf{R}(t_1,t_2) = \mathbf{R}_0 + t_1\mathbf{v}_1 + t_2\mathbf{v}_2$$

The *equations for the plane* can be written:

$$x = x_0 + t_1a_1 + t_2b_1$$
$$y = x_0 + t_1a_2 + t_2b_2$$
$$z = x_0 + t_1a_3 + t_2b_3$$

where $\mathbf{v}_1 = a_1\mathbf{i} + a_2\mathbf{j} + a_3\mathbf{k}$ and $\mathbf{v}_2 = b_1\mathbf{i} + b_2\mathbf{j} + b_3\mathbf{k}$.

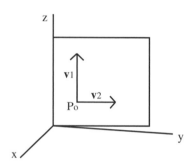

• To *parameterize a surface* in three dimensions such as a *cylinder*, first remember that a circle in two dimensions is described using $x = \cos t$, $y = \sin t$. If the circle is on an XY plane, then the z-dimension is zero and the equations become $x = \cos t$, $y = \sin t$, $z = 0$. If z and t are allowed to vary, then many circles along the Z-axis can exist.

Therefore, $x = \cos t$, $y = \sin t$, and $z = z$ can describe many circles along z. Using position vectors:

$$\mathbf{R} = (x)\mathbf{i} + (y)\mathbf{j} + (z)\mathbf{k}$$
$$\mathbf{R} = (\cos t)\mathbf{i} + (\sin t)\mathbf{j} + (z)\mathbf{k}$$

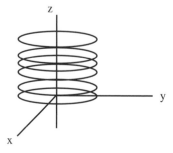

- A *cylinder* can be described by: $x^2 + y^2 = r^2$, $-1 \leq z \leq 1$, where r is the radius and the height in the z-direction is 2 with the cylinder at origin. The cylinder can be parameterized using parameters u and v, where $x = r \cos u$, $y = r \sin u$, and $z = v$.

Parameters u and v vary in the uv plane as:

$0 \leq u \leq 2\pi$ and $-1 \leq v \leq 1$

The parametric representation is:

$\mathbf{R}(u,v) = r \cos u \, \mathbf{i} + r \sin u \, \mathbf{j} + v\mathbf{k}$

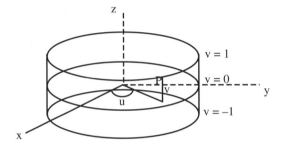

where there is a point P on the cylinder that corresponds to a (u,v) value.

- A *cone* given by: $\sqrt{x^2 + y^2} = z$, $0 \leq z \leq h$ can be parameterized using parameters u and v, where:

$x = u \cos v$, $y = u \sin v$, and $z = u$.

Parameters u and v vary in the uv plane as:

$0 \leq u \leq h$ and $0 \leq v \leq 2\pi$

The parametric representation is:

$\mathbf{R}(u,v) = u \cos v \, \mathbf{i} + u \sin v \, \mathbf{j} + u\mathbf{k}$

- A *sphere* can be parameterized using spherical coordinates. In three dimensions, spherical coordinates are expressed in terms of ρ, θ, and ϕ, where ρ can range from 0 to ∞, θ can range from 0 to 2π, and ϕ can range from 0 to π. In *spherical coordinates*, the ρ component is measured from the origin, the θ component measures the distance around the Z-axis, and the ϕ component measures down from the Z-axis and is referred to as the polar angle. The coordinates can be defined in terms of Cartesian coordinates, x, y, and z:

$$x = \rho \cos \theta \sin \phi$$
$$y = \rho \sin \theta \sin \phi$$
$$z = \rho \cos \phi$$
$$\rho = \sqrt{x^2 + y^2 + z^2}$$

The parameters of a sphere centered at origin are:

$$x = \cos \theta \sin \phi$$
$$y = \sin \theta \sin \phi$$
$$z = \cos \phi$$

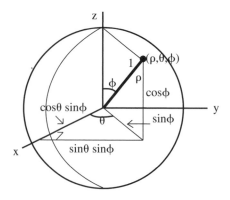

The equations can be written:

$$\mathbf{R}(\theta,\phi) = (\cos \theta \sin \phi)\mathbf{i} + (\sin \theta \sin \phi)\mathbf{j} + (\cos \phi)\mathbf{k}$$

For example, if point P is centered at (2,2,2) and the sphere has a radius of 2, then $\mathbf{R}_0 = 2\mathbf{i} + 2\mathbf{j} + 2\mathbf{k}$ and $\mathbf{R}(\theta,\phi)$ is multiplied by 2 to expand the radius to 2. The equation becomes:

$$\mathbf{R}(\theta,\phi) = 2\mathbf{i} + 2\mathbf{j} + 2\mathbf{k} + (2\cos \theta \sin \phi)\mathbf{i} + (2\sin \theta \sin \phi)\mathbf{j} + (2\cos \phi)\mathbf{k}$$
$$= (2 + 2\cos \theta \sin \phi)\mathbf{i} + (2 + 2\sin \theta \sin \phi)\mathbf{j} + (2 + 2\cos \phi)\mathbf{k}$$

Or

$$x = 2 + 2\cos \theta \sin \phi$$
$$y = 2 + 2\sin \theta \sin \phi$$
$$z = 2 + 2\cos \phi$$

- The *circumference* of a circle can be written in terms of parametric equations: $x = \cos t$, $y = \sin t$, $0 \le t \le 2\pi$

The *length of the curve* or distance around the circle is:

$$\int_0^{2\pi} \sqrt{\left(\frac{dx}{dt}\right)^2 + \left(\frac{dy}{dt}\right)^2}\, dt = \int_0^{2\pi} \sqrt{(-\sin t)^2 + (\cos t)^2}\, dt = 2\pi$$

Note $(-\sin t)^2 = (\sin t)^2$

- The *length of a quarter-circle* can be written in terms of parametric equations: $x = \cos t$, $y = \sin t$, $0 \le t \le \pi/2$

The length of the curve or distance along the arc is:

$$\int_0^{\pi/2} \sqrt{\left(\frac{dx}{dt}\right)^2 + \left(\frac{dy}{dt}\right)^2}\, dt = \int_0^{\pi/2} \sqrt{(\sin t)^2 + (\cos t)^2}\, dt = \int_0^{\pi/2} dt = \pi/2$$

- The *length of a parametric curve* can also be written by a sum of incremental lines Δs, where:

$$(\Delta s)^2 = (\Delta x)^2 + (\Delta y)^2 \text{ and } x = x(t), y = y(t)$$

Therefore:

$$\Delta s \approx \sqrt{(\Delta x)^2 + (\Delta y)^2}$$

In integral form:

$$\int ds = \int (ds/dt)\, dt = \int \sqrt{(dx/dt)^2 + (dy/dt)^2}\, dt$$

- A *unit circle* can be represented implicitly, explicitly, or parametrically as follows:

  implicitly: $x^2 + y^2 = 1$

  explicitly: $y = \sqrt{1-x^2}$ and $y = -\sqrt{1-x^2}$

  parametrically: $x = \cos t$, $y = \sin t$, $0 \le t \le 2\pi$

- A *curve* can be represented implicitly, explicitly, or parametrically in an XY plane as: (a.) *implicitly* by an equation in x and y, or f(x,y); (b.) *explicitly* by equations for y in terms of x or x in terms of y, $y = f(x)$ or $x = g(y)$; or (c.) *parametrically* by a pair of equations for x and y in terms of a third variable or parameter, $x = f(t)$ and $y = g(t)$.

5.9 Motion, Velocity, and Acceleration

• This section includes motion of a particle in a line, in a plane, on a circle, along a curve, along a cycloid path, and at a constant velocity. Also included in this section is representing velocity and acceleration using the position vector and parametric form, and general expressions for velocity and acceleration.

• The *motion of an object in a straight line* can be described using a single variable function f(t). The motion of an object along a curve or surface in two or three dimensions can be described using all the coordinates of the curve or surface, for example x(t), y(t), and z(t). *Parametric equations* can be used to describe this motion.

• For uniform *motion in a straight line*, the speed, direction, and velocity remain constant, and the equation of the line can be written:

$$\mathbf{R}(t) = \mathbf{R}_0 + t\mathbf{v}$$

where $\mathbf{R}_0 = x_0\mathbf{i} + y_0\mathbf{j} + z_0\mathbf{k}$, is the starting point.

The velocity is $\mathbf{V} = v_1\mathbf{i} + v_2\mathbf{j} + v_3\mathbf{k}$.

The speed is $|\mathbf{V}| = \sqrt{v_1^2 + v_2^2 + v_3^2}$.

The *direction of the line* is the *unit vector* $\mathbf{V}/|\mathbf{V}|$.

Separating the x, y, and z components gives the *equation of the line* in terms of parameter t:

$$x = x_0 + t(x_1 - x_0) = x_0 + tv_1$$
$$y = y_0 + t(y_1 - y_0) = y_0 + tv_2$$
$$z = z_0 + t(z_1 - z_0) = z_0 + tv_3$$

By rearranging:

$$t = (x - x_0)/v_1 = (y - y_0)/v_2 = (z - z_0)/v_3$$

which gives the *equation of the line* without parameter t:

$$\frac{x - x_0}{v_1} = \frac{y - y_0}{v_2} = \frac{z - z_0}{v_3}$$

• *Motion of a particle in a plane* can be described using two *parametric equations*, x = f(t) for horizontal motion along the x-coordinate and y = g(t) for vertical motion along the y-coordinate. The parameter is t for time such that at time t the particle is at point (f(t), g(t)).

• For example, the *motion of a particle on a circle* in a plane can be described using the parametric equations x = cos t and y = sin t, where t represents time.

A circle with radius 1 can be expressed generally as:

$$x^2 + y^2 = 1$$

which can be written using the parameters:

x = cos t and y = sin t as:
$$\cos^2 t + \sin^2 t = 1$$

If a particle is moving at a uniform speed, it will travel around the circle in 2π units of time. As the particle travels around the circle, its motion can be reflected onto the X- and Y-axes such that it goes from −1 through zero to +1 in both x and y directions (for a unit circle).

If this particle is moving in a counterclockwise direction uniformly, at different t values x and y are:

at t = 0:   x = 1, y = 0
at t = π/2:   x = 0, y = 1
at t = π: x = −1, y = 0
at t = 3π/2:   x = 0, y = −1
at t = 2π: x = 1, y = 0

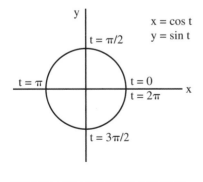

x = cos t
y = sin t

2 | t = π/2

1,3,5 | t = 0 = π = 2π

4 | t = 3π/2

Y-axis reflection moving
through t values
of t = 0 to t = 2π

X-axis reflection
moving through points 1 to 5

• *Parametric equations* can be used to describe the motion of a particle *moving on a curve* as well as to describe the curve itself. A curve is generally *parameterized* from one end to the other without retracing. To parameterize function y = f(x), substitute the parameter t for x, x = t, y = f(t), where parameter t may or may not represent time.

• **Example:** The parameterization of a half-circle from 0 to π is: x = cos t, y = sin t, $0 \leq t \leq \pi$.

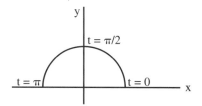

• **Example:** A *projectile*, which is defined to experience the force of gravity and no frictional forces, can be described using the x (horizontal) component, the y (vertical) component, and time t.

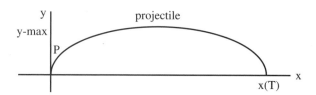

The initial position is x(t) = 0 and y(t) = 0, or in some cases a starting height y = h is specified. The initial velocity in the x direction is ($v_0 \cos \theta$) and in the y direction is ($v_0 \sin \theta$), where v_0 is the speed and θ is the angle the projectile makes with the horizontal axis. The force (or acceleration) of gravity is in the y-direction and is given by: $d^2y/dt^2 = -g$, where gravity affects the upward component of velocity so that it decreases by (−gt). The horizontal component of velocity remains constant. Therefore, the x and y components of velocity are:

$$v_x = dx/dt = v_0 \cos \theta$$
$$v_y = dy/dt = v_0 \sin \theta - gt$$

The distance along the X-axis $x(t)$ increases with time and the height along the Y-axis $y(t)$ increases, then decreases. The *distance traveled* or path of the projectile is obtained by integrating the velocity components with respect to time resulting in:

$x(t) = (v_0 \cos \theta)t$

$y(t) = (v_0 \sin \theta)t - gt^2/2$

The *maximum height* occurs where $dy/dt = 0$. When

$dy/dt = 0$, then $v_0 \sin \theta = gt$.

Solve $v_0 \sin \theta = gt$ for t:

$t = (v_0 \sin \theta)/g$

Substitute t into $y(t)$ to obtain y_{max}:

$y(t)$ at $y_{max} = (v_0 \sin \theta)(v_0 \sin \theta)/g - g((v_0 \sin \theta)/g)^2/2$

$= (v_0 \sin \theta)^2/g - (v_0 \sin \theta)^2/2g$

$= [(v_0 \sin \theta)^2/g] (1 - 1/2) =$

$y_{max} = (v_0 \sin \theta)^2/2g$

where y_{max} occurs at one-half of the time the projectile is in the air.

The *horizontal distance* $x(T)$ the projectile travels occurs when $y = 0$ and time $= T$, where T is the total time in the air. Therefore, at time $= T$:

$(v_0 \sin \theta)T = gT^2/2$

The total time T is:

$T = (2v_0 \sin \theta)/g$

The *total distance* $x(t)$ at $t = T$ is:

$x(T) = (v_0 \cos \theta)T = (v_0 \cos \theta)(2 v_0 \sin \theta)/g = (v_0^2 \sin 2\theta)/g$

Remember, $2 \sin x \cos x = \sin 2x$.

• **Example:** A *cycloid* is described by the path of a point on the perimeter of a circle as it is rolled along a line or surface. If the circle has radius r, and point P begins at the bottom at $x = 0$, then if it is rolled along the X-axis it makes a complete revolution at $x = 2\pi r$.

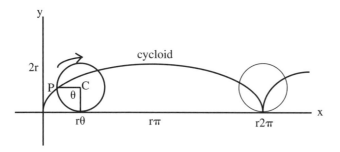

The parameter θ represents the angle through which the circle revolves. The circle rolls a distance of rθ along the X-axis and its center is at y = r and x = rθ. At θ = 0 the point is at x = 0, y = 0 and at θ = 2π, the point is at x = 2πr, y = 0. The segment between the center and the point is taken into account in measurements by subtracting (r sin θ) from x and (r cos θ) from y. Therefore:

$$x = r\theta - r\sin\theta = r(\theta - \sin\theta)$$
$$y = r - r\cos\theta = r(1 - \cos\theta)$$

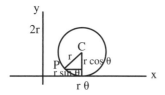

The *slope of the cycloid* (dy/dx) is derived using the chain rule:

$$\frac{dy}{dx} = \frac{dy/d\theta}{dx/d\theta} = \frac{r\sin\theta}{r(1-\cos\theta)}$$

At θ = 0, the slope is infinite where the point moves straight up.

The *arc length of a cycloid* can be found by integrating ds from 0 to 2π on the X-axis:

$$\int ds = \int_0^{2\pi} \sqrt{(dx/d\theta)^2 + (dy/d\theta)^2}\, d\theta$$

$$\int_0^{2\pi} r\sqrt{(1-\cos\theta)^2 + (\sin\theta)^2}\, d\theta$$

The *area of a cycloid* can be found by integrating y dx from 0 to 2π on the X-axis:

$$\int y\,dx = \int_0^{2\pi} r(1-\cos\theta)r(1-\cos\theta)\,d\theta$$

where $y = r(1 - \cos\theta)$ and $dx = r(1 - \cos\theta)d\theta$.

• A particle *moving at a constant velocity* can be represented using the *position vector* of the particle at time t, which is **R**(t). Then the *displacement vector* between positions at time t and time $t + \Delta t$ can be written generally as:

$$d\mathbf{R} = \mathbf{R}(t + \Delta t) - \mathbf{R}(t)$$

Therefore, $\mathbf{V}(t) \approx \Delta\mathbf{R}/\Delta t$.

The *velocity of a moving particle* in terms of the position vector **R**(t) at time t can be represented using:

$$\mathbf{V}(t) = \lim_{\Delta t \to 0}[\mathbf{R}(t + \Delta t) - \mathbf{R}(t)/\Delta t]$$

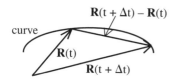

In an XY plane the vectors **R**, $\Delta\mathbf{R}$, and $\mathbf{V} = d\mathbf{R}/dt$ can be depicted as:

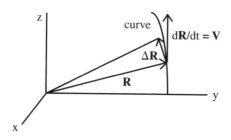

where $\Delta\mathbf{R}$ is the change between two points on the curve.

• In three dimensions the *velocity vector can be represented* using the components of the *position vector **R** in parametric form*, where x = f(t), y = g(t), z = h(t), and:

$$\mathbf{R} = f(t)\mathbf{i} + g(t)\mathbf{j} + h(t)\mathbf{k}, \text{ as:}$$

$$\mathbf{V}(t) = \lim_{\Delta t \to 0}[\mathbf{R}(t + \Delta t) - \mathbf{R}(t)/\Delta t] =$$

$$\lim_{\Delta t \to 0}[\frac{f(t+\Delta t) - f(t)}{\Delta t}\mathbf{i} + \frac{g(t+\Delta t) - g(t)}{\Delta t}\mathbf{j} + \frac{h(t+\Delta t) - h(t)}{\Delta t}\mathbf{k}]$$

$$= \frac{df(t)}{dt}\mathbf{i} + \frac{dg(t)}{dt}\mathbf{j} + \frac{dh(t)}{dt}\mathbf{k}$$

Because x = f(t), y = g(t), and z = h(t), the components of the *velocity vector* **V** in parametric form for a particle moving in three-dimensional space can be written:

$$\mathbf{V}(t) = \frac{dx}{dt}\mathbf{i} + \frac{dy}{dt}\mathbf{j} + \frac{dz}{dt}\mathbf{k}$$

• The *acceleration vector* **A** of a particle moving with a velocity **V**(t) at time t can be represented in general as:

$$\mathbf{A}(t) = \lim_{\Delta t \to 0}[\Delta \mathbf{V}/\Delta t] = \lim_{\Delta t \to 0}[\mathbf{V}(t + \Delta t) - \mathbf{V}(t)/\Delta t]$$

For a particle moving in three-dimensional space at time t the *acceleration* is given by:

$$\mathbf{A}(t) = d\mathbf{V}/dt = d^2\mathbf{R}/dt^2 = \frac{dx^2}{dt^2}\mathbf{i} + \frac{dy^2}{dt^2}\mathbf{j} + \frac{dz^2}{dt^2}\mathbf{k}$$

• If a particle is moving around a circle with radius R at a constant speed of $|\mathbf{V}|$, then its *acceleration vector* is pointing toward the center. In this case of uniform motion the velocity vector changes in direction but not magnitude.

• Consider a position vector **R** at some point P on a curve in two dimensions represented with parametric equations x = f(t) and y = g(t) where t is a scalar and **R** is given by:

$$\mathbf{R} = f(t)\mathbf{i} + g(t)\mathbf{j}$$

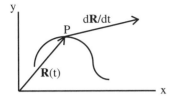

In three dimensions, the *position vector* **R** *for point* P = (x,y,z) on the curve can be written using parametric equations x = f(t), y = g(t), and z = h(t):

R = f(t)**i** + g(t)**j** + h(t)**k**

Because **R**(t) is the position vector for point P and d**R**/dt = **V**(t) is the *velocity vector at point* P, then the *acceleration vector at* P is d²**R**/dt² = d**V**/dt = **A**(t).

The *derivative of vector* **R** is the *vector tangent* to the curve in two dimensions at point P pointing in the direction of motion and represents *velocity* **V** of a moving particle at point P, providing d**R**/dt ≠ 0. For a moving particle, the direction of **V** is the direction of motion and the magnitude of **V** is the speed of the particle. The *velocity* in two dimensions is given by:

$$\mathbf{V}(t) = \frac{d\mathbf{R}}{dt} = \frac{df}{dt}\mathbf{i} + \frac{dg}{dt}\mathbf{j}$$

The *magnitude* (or length) of vector d**R**/dt in two dimensions is:

$$\left|\frac{d\mathbf{R}}{dt}\right| = \sqrt{\left(\frac{df}{dt}\right)^2 + \left(\frac{dg}{dt}\right)^2}$$

• Similarly, in three dimensions, the derivative of the vector **R** tangent to the curve at point P written is **V**(t):

$$\mathbf{V}(t) = \frac{d\mathbf{R}}{dt} = \frac{df}{dt}\mathbf{i} + \frac{dg}{dt}\mathbf{j} + \frac{dh}{dt}\mathbf{k}$$

The *magnitude* (or length) of vector d**R**/dt in three dimensions is:

$$|\mathbf{V}(t)| = \left|\frac{d\mathbf{R}}{dt}\right| = \sqrt{\left(\frac{df}{dt}\right)^2 + \left(\frac{dg}{dt}\right)^2 + \left(\frac{dh}{dt}\right)^2}$$

The magnitude of the velocity vector also represents the *speed of a particle traveling along the curve*, where

x = f(t), y = g(t), and z = h(t):

$$|V| = \sqrt{\left(\frac{dx}{dt}\right)^2 + \left(\frac{dy}{dt}\right)^2 + \left(\frac{dz}{dt}\right)^2}$$

Remember that velocity is a vector with magnitude and direction, while speed has only magnitude.

The *unit tangent vector* in three dimensions can be written:

$$T = \frac{V}{|V|} = \frac{\dfrac{df}{dt}i + \dfrac{dg}{dt}j + \dfrac{dh}{dt}k}{\sqrt{\left(\dfrac{df}{dt}\right)^2 + \left(\dfrac{dg}{dt}\right)^2 + \left(\dfrac{dh}{dt}\right)^2}}$$

• Note that by integrating $|V|$, the *distance traveled of a particle along the curve* and the *length of the curve* can be obtained:

$$\text{Distance Traveled} = \int_a^b |V(t)|\, dt$$

$$\text{Length of Curve} = \int_a^b |V|\, dt$$

where the curve is in a plane and a ≤ t ≤ b.

• The unit *tangent vector* **T** points in the *direction of motion* and depends on the shape of the curve. The *curvature* is sometimes referred to as κ and is equal to the change in direction divided by the change in position, or the rate of turning:

κ = $|dT/ds|$ = $|(dT/dt)/(ds/dt)|$ = $|d^2R/ds^2|$

where **T** and d**T**/dt are perpendicular to each other.

The unit *normal vector* **N** describes the *direction of turning* and is perpendicular to **T**. Therefore, the unit normal vector **N** coincides with d**T**/dt and is perpendicular to **T**:

$$N = \frac{dT/ds}{|dT/ds|} = \frac{dT/ds}{\kappa} = \frac{(dT/ds)}{|dT/dt|/(ds/dt)}$$

$$= \frac{(dT/ds)(ds/dt)}{|dT/dt|} = \frac{dT/dt}{|dT/dt|}$$

The *acceleration along a curve* has a straight line component and a curvature component, and can be represented in several forms:

$A = (d^2s/dt^2)T + \kappa(ds/dt)^2N$
$= (d^2s/dt^2)T + \kappa|V|^2N$
$= (d^2s/dt^2)T + |(dT/ds)|(ds/dt)^2(dT/ds)/|dT/ds|$
$= (d^2s/dt^2)T + (ds/dt)^2(dT/ds)$
$= (d^2s/dt^2)T + (ds/dt)(dT/dt)$

• Note, for example, that the vector $1i + 2j + 3k$ is constant, and the vector $t^2i + t^2j + t^2k$ is moving, where t is the parameter time.

• *Motion in a plane in polar coordinates* can be summarized using the position vector **R**, which points out from the origin. For circular motion such as a planet moving in a plane, the unit vector along **R** is:

$u_r = R/r = (xi + yj)/r = \cos\theta\, i + \sin\theta\, j$

where r is the radius.

The unit vector u_θ is perpendicular to u_r and is around the center:

$u_\theta = -\sin\theta\, i + \cos\theta\, j$

If r varies, u_r and u_θ are unaffected. However, if θ varies

u_r and u_θ are:
$du_r/d\theta = -\sin\theta\, i + \cos\theta\, j = u_\theta$
$du_\theta/d\theta = -\cos\theta\, i - \sin\theta\, j = -u_r$

The *velocity* $V = dR/dt$ can be determined using: $u_r = R/r$, or $R = ru_r$. Using the chain rule for $du_r/dt = (du_r/d\theta)(d\theta/dt) = (d\theta/dt)u_\theta$, then velocity becomes:

$V = dR/dt = d(ru_r)/dt = u_r(dr/dt) + ru_\theta(d\theta/dt)$

where the outward speed is dr/dt and the circular speed is $d\theta/dt$.

Using the chain rule and the equation for **V**, the *acceleration* for motion in polar coordinates is:

$$\mathbf{A} = \frac{d\mathbf{V}}{dt} = \frac{d^2r}{dt^2}\mathbf{u}_r + \frac{dr}{dt}\frac{d\mathbf{u}_r}{dt} + \frac{dr}{dt}\frac{d\theta}{dt}\mathbf{u}_\theta + r\frac{d^2\theta}{dt^2}\mathbf{u}_\theta + r\frac{d\theta}{dt}\frac{d\mathbf{u}_\theta}{dt}$$

Using:

$(d\mathbf{u}_r/d\theta)(d\theta/dt) = (d\theta/dt)\mathbf{u}_\theta = (d\mathbf{u}_r/dt)$ and

$(d\mathbf{u}_\theta/d\theta)(d\theta/dt) = -(d\theta/dt)\mathbf{u}_r = -(d\mathbf{u}_\theta/dt)$

$$\mathbf{A} = \frac{d^2r}{dt^2}\mathbf{u}_r + \frac{dr}{dt}\frac{d\theta}{dt}\mathbf{u}_\theta + \frac{dr}{dt}\frac{d\theta}{dt}\mathbf{u}_\theta + r\frac{d^2\theta}{dt^2}\mathbf{u}_\theta - r\frac{d\theta}{dt}\frac{d\theta}{dt}\mathbf{u}_r$$

$$= \left(\frac{d^2r}{dt^2} - r\left(\frac{d\theta}{dt}\right)^2\right)\mathbf{u}_r + \left(r\frac{d^2\theta}{dt^2} + 2\frac{dr}{dt}\frac{d\theta}{dt}\right)\mathbf{u}_\theta$$

• In general, expressions for *velocity* and *acceleration* of a particle moving along a curve (*curvilinear motion*) are more complicated than for a particle moving in a straight line. The equation of the curve can be given in parametric form as $x = f(t)$ and $y = g(t)$, where t represents time. Velocity **V** is a *vector tangent* to the curve and has an x and a y component and is expressed in terms of magnitude (speed) and direction.

• *Velocity* **V** can be defined in terms of x and y components in two dimensions and written as follows:

The x component of velocity is $v_x = dx/dt$.

The y component of velocity is $v_y = dy/dt$.

The *magnitude* (speed) of **V** is $|\mathbf{V}| = \sqrt{v_x^2 + v_y^2}$.

The *direction* of **V** is $\tan\emptyset = v_y/v_x = dy/dx$.

• *Acceleration* is given for the x and y components in two dimensions and can be written as follows:

The x component of acceleration is $a_x = dv_x/dt = d^2x/dt^2$.

The y component of acceleration is $a_y = dv_y/dt = d^2y/dt^2$.

The *magnitude* of **A** is $|\mathbf{A}| = \sqrt{a_x^2 + a_y^2}$.

The *direction* of **A** is $\tan\emptyset = a_y/a_x$.

The *acceleration vector* can be expressed in a tangent component and a normal component to the curve, which are perpendicular to each other.

The *tangent component of acceleration* $a_T = \dfrac{v_x a_x + v_y a_y}{|V|}$.

The *normal component of acceleration* $a_N = \dfrac{v_x a_y - v_y a_x}{|V|}$.

Chapter

6

Partial Derivatives

6.1 Partial Derivatives: Representation and Evaluation

• This section provides the definition and notation for partial derivatives and evaluating first, second, and third partial derivatives.

• Derivatives of functions containing one variable generally represent aspects of curves, tangent lines, and rates of change. *Partial derivatives* involve functions of more than one variable and typically represent aspects of surfaces, tangent planes, and rates of change. In these functions each variable can change independently of the other variable(s), thus affecting change in the function as a whole.

• In a graph of $y = f(x)$, x is the *independent variable*, y is the *dependent variable*, and points that satisfy $y = f(x)$ fall on the curve described by $y = f(x)$. Similarly, in a graph of $z = f(x,y)$, x and y are the *independent variables*, z is the *dependent variable*, and points that satisfy $z = f(x,y)$ fall on the surface described by $z = f(x,y)$.

• Partial derivatives are represented using ∂ rather than d. *Notation for single partial derivatives* includes: $(\partial f/\partial x)$, f_x, $(\partial f/\partial y)$, f_y. Notation for *second and third partial derivatives* includes: $(\partial^2 f/\partial x^2)$, f_{xx}, $(\partial^3 f/\partial x^3)$, $(\partial^3 f/\partial xxy)$, f_{xxy}.

• To differentiate or solve a partial derivative of function $z = f(x,y)$, hold variable y constant while differentiating variable x, then hold x constant while differentiating y. The variable that is being held constant is treated as a constant during each differentiation. For each small change in variable x or variable y, the function z will change.

• For a change in x or Δx, z changes and the *definition of the partial derivative* for $z = f(x,y)$ becomes:

$$\frac{\partial z}{\partial x} = \frac{\partial f}{\partial x} = \lim_{\Delta x \to 0} \frac{f(x + \Delta x, y) - f(x,y)}{\Delta x}$$

For a change in y or Δy, z changes and the *definition of the partial derivative* for $z = f(x,y)$ becomes:

$$\frac{\partial z}{\partial y} = \frac{\partial f}{\partial y} = \lim_{\Delta y \to 0} \frac{f(x, y + \Delta y) - f(x,y)}{\Delta y}$$

The total partial derivative of $z = f(x,y)$, when both x and y change, is:

$$dz = \frac{\partial f}{\partial x} dx + \frac{\partial f}{\partial y} dy$$

• For example, if $z = f(x,y) = x^2 + y^2 + x^2y^2$, then $(\partial f/\partial x) = 2x + 2xy^2$ and $(\partial f/\partial y) = 2y + 2yx^2$.

• If $z = f(x,y)$ and x and y each depend on time t, then the total partial derivative can be written:

$$\frac{dz}{dt} = \frac{\partial f}{\partial x} \frac{dx}{dt} + \frac{\partial f}{\partial y} \frac{dy}{dt}$$

- If $z = f(x,y)$ and x and y each depend on two variables u and v, such that $x = x(u,v)$ and $y = y(u,v)$, then the total partial derivative is written as two derivatives:

$$\frac{\partial f}{\partial u} = \frac{\partial f}{\partial x}\frac{\partial x}{\partial u} + \frac{\partial f}{\partial y}\frac{\partial y}{\partial u}$$

$$\frac{\partial f}{\partial v} = \frac{\partial f}{\partial x}\frac{\partial x}{\partial v} + \frac{\partial f}{\partial y}\frac{\partial y}{\partial v}$$

- It is possible to *evaluate the partial derivatives*, or slopes, $(\partial z/\partial x)$ and $(\partial z/\partial y)$ of function $z = f(x,y)$ at a point. If $z = x^2/y^2$, then evaluate $(\partial z/\partial x)$ and $(\partial z/\partial y)$ at point (3,2). To evaluate $(\partial z/\partial x)$ hold y constant at 2, differentiate with respect to x, then substitute 3 into the resulting expression:

$$x^2/y^2 = x^2/2^2$$
$$(\partial z/\partial x) = 2x/4$$
$$\text{at } x = 3, 6/4 = 3/2$$

To evaluate $(\partial z/\partial y)$, hold x constant at 3, differentiate with respect to y, then substitute 2 into the resulting expression:

$$x^2/y^2 = 3^2/y^2$$
$$(\partial z/\partial y) = (9)(-2y^{-2-1}) = -18/y^3$$
$$\text{at } y = 2, -18/2^3 = -18/8 = -9/4$$

Alternatively, $(\partial z/\partial x)$ and $(\partial z/\partial y)$ can be determined first, then the point (3,2) substituted into the two resulting equations:

$$(\partial z/\partial x) = 2x/y^2 = 6/4 = 3/2$$
$$(\partial z/\partial y) = -2x^2/y^3 = -18/8 = -9/4$$

- To evaluate partial derivatives with more than two variables, differentiate with respect to one variable at a time while treating the other variables as constants. For example, given a *partial derivative with three variables*, $w = f(x,y,z) = x^2y^2/z$, find $(\partial w/\partial x)$, $(\partial w/\partial y)$, and $(\partial w/\partial z)$:

$$(\partial w/\partial x) = 2xy^2/z$$
$$(\partial w/\partial y) = 2x^2y/z$$
$$(\partial w/\partial z) = -x^2y^2/z^2$$

• *Second partial derivatives* of a function are represented using the following *notation*: $(\partial^2 f/\partial x^2)$, $(\partial^2 f/\partial y^2)$, $(\partial^2 f/\partial xy)$, $(\partial^2 f/\partial yx)$, $(\partial/\partial x)(\partial f/\partial x)$, $(\partial/\partial x)(\partial f/\partial y)$, $(\partial/\partial y)(\partial f/\partial y)$, or equivalently, f_{xx}, f_{yy}, f_{xy}, f_{yx}, $(f_x)_x$, $(f_y)_y$, $(f_x)_y$, where f_{xx} and $(f_x)_x$ are equivalent and f_{xy} and f_{yx} are generally equivalent because the order of differentiation for most functions doesn't matter. More specifically, if $f_{xy}(x_1,y_1)$ and $f_{yx}(x_1,y_1)$ are both continuous at point (x_1,y_1), then $f_{xy}(x_1,y_1)$ and $f_{yx}(x_1,y_1)$ are equivalent.

• For example, if $z = e^x \cos y$, find $(\partial^2 z/\partial x^2)$, $(\partial^2 z/\partial y^2)$ and $(\partial^2 z/\partial xy)$:

$(\partial^2 z/\partial x^2) = (\partial/\partial x)\, e^x \cos y = e^x \cos y$

$(\partial^2 z/\partial y^2) = -(\partial/\partial y)\, e^x \sin y = -e^x \cos y$

$(\partial^2 z/\partial xy) = (\partial/\partial y)\, e^x \cos y = -e^x \sin y$

• *Notation* for *third derivatives* includes: $(\partial^3 f/\partial x^3)$, $(\partial^3 f/\partial y^3)$, $(\partial^3 f/\partial xxy)$, $(\partial^3 f/\partial yyx)$, or equivalently, f_{xxx}, f_{yyy}, f_{xxy}, f_{yyx}.

6.2 The Chain Rule

• This section presents the chain rule for partial derivatives and applying it to $f(g(x,y))$, $f(x(t),y(t))$ and $f(x(u,v),g(u,v))$.

• The *chain rule applies to partial derivatives* of more complicated functions just as it does with ordinary derivatives. The chain rule provides a means to differentiate *composite functions* and, therefore, is used for differentiating functions of functions. In an ordinary derivative, a composite function has one function in another function such as $y = f(g(x))$. Similarly, in a function with more than one variable, a composite function can have more than one function substituted within a function such as, $f(g(x,y))$, $f(x(t),y(t))$, $f(x(u,v),y(u,v))$ and $z = f(g(t),h(t))$.

• Following is a summary of differentiating common forms of functions using the chain rule (it is assumed that all the derivatives are continuous):

(a.) To differentiate $y = f(g(x))$, calculate the ordinary derivatives, $(df/dx) = (df/dg)(dg/dx)$.

(b.) To differentiate $z = f(g(x,y))$, calculate $(\partial f/\partial x)$ and $(\partial f/\partial y)$:

$(\partial f/\partial x) = (df/dg)(\partial g/\partial x)$

$(\partial f/\partial y) = (df/dg)(\partial g/\partial y)$

where f depends on g, and g depends on x and y.

(c.) To differentiate $f(x(t),y(t))$, calculate (df/dt):

$df/dt = (\partial f/\partial x)(dx/dt) + (\partial f/\partial y)(dy/dt)$

where a change in t influences a change in x and y and they influence a change in f.

(d.) To differentiate $f(x(u,v),y(u,v))$, calculate $(\partial f/\partial u)$ and $(\partial f/\partial v)$:

$(\partial f/\partial u) = (\partial f/\partial x)(\partial x/\partial u) + (\partial f/\partial y)(\partial y/\partial u)$

$(\partial f/\partial v) = (\partial f/\partial x)(\partial x/\partial v) + (\partial f/\partial y)(\partial y/\partial v)$

where changes in u and v influence changes in x and y and they influence a change in f.

• For example, differentiate $z = (x + xy)^3$.

Use case (b.) above, where $z = f(g(x,y))$ to calculate $(\partial f/\partial x)$ and $(\partial f/\partial y)$, or equivalently $(\partial z/\partial x)$ and $(\partial z/\partial y)$:

$(\partial z/\partial x) = (df/dg)(\partial g/\partial x)$

$(\partial z/\partial x) = 3(x + xy)^2(\partial/\partial x)(x + xy) = 3(x + xy)^2(1 + y)$

$(\partial z/\partial y) = (df/dg)(\partial g/\partial y)$

$(\partial z/\partial y) = 3(x + xy)^2(\partial/\partial y)(x + xy) = 3x(x + xy)^2$

6.3 Representation on a Graph

• This section includes examples of functions having more than one variable in the form $f(x,y)$ and their graphs and contour diagrams.

• In a graph of a one-variable function, $y = f(x)$, (dy/dx) represents the slope of a line drawn tangent to a curve. In a graph of a two-variable function $z = f(x,y)$, (figure below), $(\partial z/\partial x)$ represents the slope of a curve sliced from the surface of $f(x,y)$ by a plane at $y =$ constant. Similarly, $(\partial z/\partial y)$ represents the slope of a curve sliced from the surface of $f(x,y)$ by a plane at $x =$ constant. In the derivative $(\partial z/\partial x)$, x is varied and y is held constant, and in the derivative $(\partial z/\partial y)$, y is varied and x is held constant.

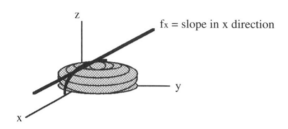

f_x = slope in x direction

• For example, consider the graph of the two-variable function $f(x,y) = z = x^2 - y^2$:

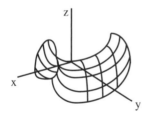

This graph forms a *saddle-shaped surface*. In this graph, there are two sets of parabolas, one set opening upward and the other set opening downward corresponding to x^2 values and $-y^2$ values. Each curve corresponds to $f(x,y)$ when x is held constant and y is varied or when y is held constant and x is varied. The partial derivatives of $f(x,y)$ are $(\partial z/\partial x) = 2x$ and $(\partial z/\partial y) = -2y$. The point where the two sets of upward and downward parabolas meet is called the saddle point. The derivatives at the saddle point are zero.

A *contour diagram* perspective of $f(x,y) = z = x^2 - y^2$ can also be depicted:

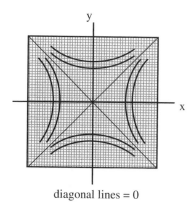

diagonal lines = 0

• A contour diagram of curved surfaces can be depicted by connecting all of the points at the same height on the surface, such that all points satisfying f(x,y) = c at a given c lie on each *contour line*. The *level curves* form loops around the maximum point(s). As the height increases the loops get smaller. In other words, level curves, or contour lines, can be seen by slicing a surface with horizontal planes. The contour line at each height h = z is represented by f(x,y) = h. By moving in a direction parallel to an axis and crossing over the contour lines, the partial derivative equals the rate of change of the value of the function on the contour lines.

• Another example of a two-variable function is a graph of
f(x,y) = z = $(x^2 + y^2)^{1/2}$:

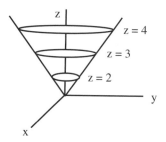

This graph has circular planes at each z value. When f(x,y) = z = (a constant), a *contour* map or diagram from the top-down perspective of z = $(x^2 + y^2)^{1/2}$ can be drawn:

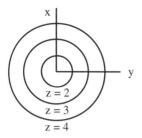

For example, if x and y equal 2, 3, and 4, then: $f(x,y) = (x^2 + y^2)^{1/2}$ is equal to $2\sqrt{2}$, $3\sqrt{2}$, and $4\sqrt{2}$ respectively. The partial derivatives of $z = (x^2 + y^2)^{1/2}$ are obtained by varying each variable while holding the other constant:

$$(\partial z/\partial x) = (1/2)2x(x^2 + y^2)^{-1/2} = x/\sqrt{x^2 + y^2}$$

$$(\partial z/\partial y) = (1/2)2y(x^2 + y^2)^{-1/2} = y/\sqrt{x^2 + y^2}$$

The partial derivatives or slopes where x and y equal 2, 3, and 4 are equal to:

$$2/2\sqrt{2} = 1/\sqrt{2}$$
$$3/3\sqrt{2} = 1/\sqrt{2}$$
$$4/4\sqrt{2} = 1/\sqrt{2}$$

which would be expected.

6.4 Local Linearity, Linear Approximations, Quadratic Approximations, and Differentials

• This section includes linear approximations, quadratic approximations, local linearity, tangent lines and planes, the normal line equation, quadratic approximations, Taylor polynomials, and differentials.

• When *calculating approximate values for complicated functions* it is sometimes possible to focus in on a small region of the graph of a function, and look at that region as if it were linear. This is sometimes referred to as a point of *local linearity*. A *tangent line* can be drawn through a point in a locally linear region and the slope of the tangent line is the derivative of the function at that point. Local linearity is used in one-variable functions $y = f(x)$ to focus in on a curve until it appears to be a straight line (which is tangent to the curve) and can be described by a linear function or *linear approximation* at that point.

• Similarly, in two-variable functions, *local linearity* can be used to focus in on a curved surface until it appears to be a flat plane (which is a *tangent plane* to the surface) and can be described by a linear function at that point. A function is differentiable at a point if it is locally linear and continuous at that point. The slope of a tangent plane at a point measures the change in the curve at that point. Tangents are used in representing *linear approximations* of functions and small changes in a function.

• When approximating values of *one-variable functions*, remembering the following two facts can be useful:

(a.) The *slope of a line drawn tangent* to a graph of a function at a point is the *derivative* of the function at that point. In other words, the *slope of the tangent* at point $(a, f(a))$ equals the derivative $f'(a)$.

(b.) The *equation for a tangent* line passing through some point $(a, f(a))$ is $y - f(a) = f'(a)(x - a)$. Equivalently, the equation for the slope of a line passing through point (x_1, y_1) is $m = (y - y_1) / (x - x_1)$, where m is the slope of the tangent line and derivative at point (x_1, y_1).

• In the graph of function $z = f(x, y)$, the *slope of a plane drawn tangent* to the curved surface through some point $(a, b, f(a, b))$ on the surface is the derivative of the function at point $(a, b, f(a, b))$. In other words, the *slope of the tangent plane* at point $(a, b, f(a, b))$ equals the *partial derivative* of the function in the x-direction at that point and the partial derivative of the function in the y-direction at that point. The slope of a tangent at a point measures the change in the surface at that point and is described by the equation for the plane drawn tangent to the surface.

• For a point on the surface of $z = f(x, y)$ at $x = x_1$, $y = y_1$, $z = z_1$, *the equation for the tangent plane* (where z is the dependent variable) in that seemingly linear region is the following linear equation:

$$z - z_1 = (\partial f/\partial x)_1 (x - x_1) + (\partial f/\partial y)_1 (y - y_1)$$

where $(\partial f/\partial x)_1$ and $(\partial f/\partial y)_1$ represent the *slopes of the tangent plane* and are evaluated at point (x_1, y_1, z_1).

Similarly for $F(x,y,z) = 0$ with three variables, at point (x_1,y_1,z_1), *the equation for the tangent plane* is:

$$(\partial F/\partial x)_1(x - x_1) + (\partial F/\partial y)_1(y - y_1) + (\partial F/\partial z)_1(z - z_1) = 0$$

Note that a two-variable function $z = f(x,y)$ represents a single *surface* and a three-variable function $F(x,y,z)$ represents a *family of level surfaces*. $F(x,y,z) = f(x,y) - z$ for one of the surfaces and $z = f(x,y)$ is the surface at $F(x,y,z) = 0$.

• For example, if a surface is given by $F(x,y,z) = x^2 + y^2 + z^2$, at point $(x_1,y_1,z_1) = (1,2,3)$ the equation for a tangent plane

$$(\partial F/\partial x)_1(x - x_1) + (\partial F/\partial y)_1(y - y_1) + (\partial F/\partial z)_1(z - z_1) = 0, \text{ is:}$$
$$(2x)_1(x - x_1) + (2y)_1(y - y_1) + (2z)_1(z - z_1) = 0$$

At point $(1,2,3)$ the equation becomes:

$$(2)(x - 1) + (4)(y - 2) + (6)(z - 3) = 0$$

• A useful relationship to the equation of a line is the *equation for a line normal* (or perpendicular) to a *tangent line* on the curve $y = f(x)$ at a given point (x_1,y_1) or $(a,f(a))$. Because the slope is m and slopes of perpendicular lines multiply to equal -1, then the normal is $-1/m$. The equation for the normal line can be written:

$$y - y_1 = (-1/m)(x - x_1) \text{ or } y - f(a) = (-1/f'(a))(x - a).$$

Similarly for function $F(x,y,z) = 0$, an *equation for a line or vector normal* to the *tangent plane* on the surface at point (x_1,y_1,z_1) is:

$$(\partial F/\partial x)_1\mathbf{i} + (\partial F/\partial y)_1\mathbf{j} + (\partial F/\partial z)_1\mathbf{k}$$

• *Quadratic approximations* are similar to linear approximations but are generally more accurate. Because the *equation for a tangent line* passing through some point $(a,f(a))$, is $y - f(a) = f'(a)(x - a)$, the *linear approximation* for the single variable function at that point is:

$$y = f(x) \approx f(a) + f'(a)(x - a)$$

By using the second-order *Taylor polynomial* this equation form becomes a *quadratic approximation* for f near a:

$$f(x) \approx f(a) + f'(a)(x - a) + (1/2)f''(a)(x - a)^2$$

- Consider a point on the surface of $z = f(x,y)$ at $x = x_1$, $y = y_1$, $z = z_1$, *the equation for the tangent plane* is:

$$z - z_1 = (\partial f/\partial x)_1(x - x_1) + (\partial f/\partial y)_1(y - y_1)$$

where $(\partial f/\partial x)_1$ and $(\partial f/\partial y)_1$ represent the slopes of the tangent plane evaluated at point (x_1, y_1, z_1).

The *linear approximation* for the function at that point is:

$$f(x,y) \approx f(x_1,y_1) + (\partial f/\partial x)_1(x - x_1) + (\partial f/\partial y)_1(y - y_1)$$

If point (x_1, y_1) is at $(0,0)$ the approximation becomes:

$$f(x,y) \approx f(0,0) + (\partial f/\partial x)_0(x) + (\partial f/\partial y)_0(y)$$

Using the second-order *Taylor polynomial* to approximate $f(x,y)$ near (x_1,y_1) gives a *quadratic approximation* for $f(x,y)$:

$$f(x,y) \approx f(x_1,y_1) + (\partial f/\partial x)_1(x - x_1) + (\partial f/\partial y)_1(y - y_1)$$
$$+ (1/2)(\partial^2 f/\partial x^2)_1(x - x_1)^2 + (\partial^2 f/\partial x \partial y)_1(x - x_1)(y - y_1)$$
$$+ (1/2)(\partial^2 f/\partial y^2)_1(y - y_1)^2$$

If point (x_1, y_1) is at $(0,0)$ the approximation becomes:

$$f(x,y) \approx f(0,0) + (\partial f/\partial x)_0(x) + (\partial f/\partial y)_0(y) + (1/2)(\partial^2 f/\partial x^2)_0(x)^2$$
$$+ (\partial^2 f/\partial x \partial y)_0(x)(y) + (1/2)(\partial^2 f/\partial y^2)_0(y)^2$$

- To solve a problem using an *approximation*, such as the second order *Taylor series*, first calculate the derivatives of the function and evaluate them at the chosen point by substituting x and y into the differentiated function, then substitute back into the series. For example, use the second-order *Taylor series* to evaluate the function

$$f(x,y) = x^2y^2 \text{ at point } (x_1,y_1) = (1,2)$$
$$f(x,y) = x^2y^2 = (1^2)(2^2) = 4$$

The derivatives are:

$$(\partial f/\partial x) = 2xy^2 = 8$$
$$(\partial f/\partial y) = 2x^2y = 4$$
$$(\partial^2 f/\partial x^2) = (\partial/\partial x)2xy^2 = 2y^2 = 8$$
$$(\partial^2 f/\partial y^2) = (\partial/\partial y)2x^2y = 2x^2 = 2$$
$$(\partial^2 f/\partial xy) = (\partial/\partial y)2xy^2 = 4xy = 8$$

Substitute into the series:

$f(x,y) \approx f(x_1,y_1) + (\partial f/\partial x)_1(x - x_1) + (\partial f/\partial y)_1(y - y_1)$
$+ (1/2)(\partial^2 f/\partial x^2)_1(x - x_1)^2 + (\partial^2 f/\partial x \partial y)_1(x - x_1)(y - y_1)$
$+ (1/2)(\partial^2 f/\partial y^2)_1(y - y_1)^2$
$f(x,y) \approx 4 + 8(x - 1) + 4(y - 2) + (1/2)8(x - 1)^2$
$+ 8(x - 1)(y - 2) + (1/2)2(y - 2)^2$
$= 4 + 8(x - 1) + 4(y - 2) + 4(x - 1)^2 + 8(x - 1)(y - 2) + (y - 2)^2$

• *Differentials* are sometimes used to assess the change in the value of a function between two points or quantities. When moving along a curve $y = f(x)$, small movements or increments along the curve are represented by Δx and Δy, and small movements along a tangent line to the curve are represented by dx and dy. The respective *differentials* are:

$\Delta y \approx (dy/dx)\Delta x$ and $dy = (dy/dx)dx$

Similarly, moving along a curved surface $z = f(x,y)$, where z is the dependent variable, small movements or increments along the surface are represented by Δx, Δy, and Δz, and small movements along a *tangent plane* to the surface are represented by dx, dy, and dz. The *total differential* reflected on the tangent plane is given using the tangent plane equation:

$z - z_1 = (\partial f/\partial x)_1(x - x_1) + (\partial f/\partial y)_1(y - y_1)$

which becomes:

$dz = (\partial z/\partial x)_1 dx + (\partial z/\partial y)_1 dy$

• For example, the differential of $z = x^2 y^2$ is given by:

$dz = 2xy^2\, dx + 2x^2 y\, dy$

• If the increments dx, dy, and dz are small enough and, therefore, the distances between $(x - x_1)$, $(y - y_1)$, and $(z - z_1)$ are small, then the *linear approximation* to the function $z = f(x,y)$ has a small *error* and is therefore a valid approximation at some point. For a *curved surface* described by $z = f(x,y)$, the *linear approximation* to the surface near point (x_1,y_1) is given by:

$z = f(x,y) \approx f(x_1,y_1) + (\partial f/\partial x)_1(x - x_1) + (\partial f/\partial y)_1(y - y_1)$

Therefore, $\Delta z \approx (\partial f/\partial x)\Delta x + (\partial f/\partial y)\Delta y$.

6.5 Directional Derivative and Gradient

• This section includes the directional derivative and the gradient and their relationship to each other, definitions and notation for the directional derivative and the gradient, magnitude of a gradient vector, and the dot product relationship between the gradient and the directional derivative.

• A surface can have slopes in all directions, not just along the axes. The *directional derivative* represents the slope of a tangent line to a surface at a point in any chosen direction. As discussed in the previous paragraphs, $(\partial z/\partial x)$ and $(\partial z/\partial y)$ represent the rate of change of surface $z = f(x,y)$ in the directions of the X-axis and Y-axis, respectively.

• A small change ds along a surface in a specified direction can be represented using a *unit vector* $\mathbf{a}$, that is pointed in the designated direction and has components a_1 and a_2 that are each pointed in x and y directions.

Therefore, $\mathbf{a} = a_1\mathbf{i} + a_2\mathbf{j}$ and dz/ds can be written:

$$dz/ds = (\partial f/\partial x)a_1 + (\partial f/\partial y)a_2$$

which is the *directional derivative* of df/ds in the direction of the unit vector $\mathbf{a}$.

• *Notation for the directional derivative* of the unit vectors $\mathbf{a}$, $\mathbf{b}$, and $\mathbf{u}$, include df/ds, dz/ds, $D_\mathbf{a}f$, $f_\mathbf{a}(x_1,y_1)$, $D_\mathbf{b}f$, $f_\mathbf{b}(x_1,y_1)$, $D_\mathbf{u}f$ and $f_\mathbf{u}(x_1,y_1)$.

• For example, using the notation $f_\mathbf{a}(x_1,y_1)$, the directional derivative for $\mathbf{a}$ is written:

$$f_\mathbf{a}(x_1,y_1) = f_x(x_1,y_1)a_1 + f_y(x_1,y_1)a_2$$

• The directional derivative can be expressed using the *difference quotient* in a direction of vector $\mathbf{a}$ at point (x_1,y_1):

$$D_\mathbf{a}f = \lim_{h\to 0} \frac{f(x_1 + ha_1,\ y_1 + ha_2) - f(x_1,y_1)}{h}$$

This equation describes a small change represented by h in f(x,y) between points (x_1,y_1) and (x_1+ha_1, y_1+ha_2) in the direction of $\mathbf{a}$. Remember from Chapter 2 that the average rate of change is represented by the quotient and the instantaneous rate of change is represented by the limit of the quotient as h approaches zero.

• The equation for the *directional derivative* can be thought of in terms of a *linear approximation* to a surface near a point. An incremental change in $z = f(x,y)$ is:

$$\Delta z \approx (\partial f/\partial x)\Delta x + (\partial f/\partial y)\Delta y$$

where Δx is the change in the a_1 direction given by ha_1, or equivalently Δsa_1, and Δy is the change in the a_2 direction given by ha_2, or equivalently Δsa_2. Therefore, the linearized curve in the direction of **a** can be written:

$$\Delta z \approx (\partial f/\partial x)a_1\Delta s + (\partial f/\partial y)a_2\Delta s$$

Rearranging:

$$\Delta z/\Delta s \approx (\partial f/\partial x)a_1 + (\partial f/\partial y)a_2$$

Taking the limit as Δs approaches zero results in:

$$dz/ds = (\partial f/\partial x)a_1 + (\partial f/\partial y)a_2$$

which is the *directional derivative* of dz/ds.

Note: The equation for the directional derivative is an important equation to remember.

• The *gradient* is a vector quantity and describes the change in a function near a point. The *gradient characterizes maximum increase and indicates the direction of maximum increase of f at a selected point.* The gradient of $f(x,y)$ is written:

$$\text{grad } f = (\partial f/\partial x)\mathbf{i} + (\partial f/\partial y)\mathbf{j}$$

For $f(x,y,z)$ the gradient becomes:

$$\text{grad } f = (\partial f/\partial x)\mathbf{i} + (\partial f/\partial y)\mathbf{j} + (\partial f/\partial z)\mathbf{k}$$

Note: The equation for the gradient is an important equation to remember.

The *components* of the vector (grad f) are:

$$(\partial f/\partial x)\mathbf{i}, \ (\partial f/\partial y)\mathbf{j}, \text{ and } (\partial f/\partial z)\mathbf{k}$$

• The *gradient* of a *scalar function* $f(x,y,z)$, is a *vector function.* The gradient of $f(x,y,z)$ at point P, where f is differentiable, indicates the *direction of maximum increase* providing (grad $f \neq 0$).

- The *gradient vector* of $z = f(x,y)$ at (x_1, y_1) is pointed in the direction of where the greatest change in $f(x,y)$ occurs, (providing f is differentiable at (x_1, y_1)). The *gradient vector lies on the surface of f(x,y) in the x-y plane* and points in the direction that the surface is rising or increasing the greatest amount. In a *contour diagram*, the gradient vector points perpendicular to the *contour lines* in the direction of greatest increase in height, which is where the contour lines are closest together.

- The *magnitude or length of the gradient vector* given by $|\text{grad } f|$ is equal to the rate of change in the direction that it is pointing (providing f is differentiable). On a *contour diagram* the gradient vector has a magnitude corresponding to the degree (or grade) of the *slope*. Because the slope is greater when the contour lines are closer together, the magnitude of the gradient vector is greater for contours that are closer together. Conversely, because the slope is less when the contour lines are more separated, the magnitude of the gradient vector is smaller for contours that are farther apart.

- *Notation for grad* is ∇, which is also called *"del"* and is a vector that is an *operator* because its components are operations rather than numbers.

$$\nabla = (\partial/\partial x)\mathbf{i} + (\partial/\partial y)\mathbf{j} + (\partial/\partial z)\mathbf{k}$$

Therefore, grad $f = \nabla f = (\partial f/\partial x)\mathbf{i} + (\partial f/\partial y)\mathbf{j} + (\partial f/\partial z)\mathbf{k}$.

Notation, for the gradient of f includes:

grad $f(x,y,z) = (\partial f/\partial x)\mathbf{i} + (\partial f/\partial y)\mathbf{j} + (\partial f/\partial z)\mathbf{k}$
grad $f(x_1,y_1,z_1) = f_x(x_1,y_1,z_1)\mathbf{i} + f_y(x_1,y_1,z_1)\mathbf{j} + f_z(x_1,y_1,z_1)\mathbf{k}$

- **Example:** If $f = 3x + 2yz - 6y^2$, what is grad f?

grad $f = \nabla f = (3)\mathbf{i} + (2z - 12y)\mathbf{j} + (2y)\mathbf{k}$

- If the *directional derivative* of $f(x,y)$ at (x_1, y_1) is zero in all directions, then the gradient vector is a *zero vector*.

- The gradient can be evaluated outside the context of a coordinate system by remembering that the direction of (grad f) is where the directional derivative df/ds is greatest and the *length* $|\text{grad } f|$ is the greatest slope.

• The *dot product of a gradient vector* at point (x_1,y_1) with the *unit vector* **a** is equal to the *directional derivative* $f_a(x_1,y_1)$ pointing in the direction of **a** at point (x_1,y_1). Therefore:

$$\text{grad } f(x_1,y_1) \bullet \mathbf{a} = f_a(x_1,y_1) = ((\partial f/\partial x)\mathbf{i} + (\partial f/\partial y)\mathbf{j}) \bullet (a_1\mathbf{i} + a_2\mathbf{j})$$
$$= (\partial f/\partial x)a_1 + (\partial f/\partial y)a_2$$
$$= |\text{grad } f(x_1,y_1)| \cos \theta = |((\partial f/\partial x)\mathbf{i} + (\partial f/\partial y)\mathbf{j})| \cos \theta$$

where $\mathbf{a} = a_1\mathbf{i} + a_2\mathbf{j}$.

Remember the dot product of two vectors is:

$$\mathbf{A} \bullet \mathbf{B} = |\mathbf{A}||\mathbf{B}| \cos \theta$$

where $|\mathbf{A}|$ and $|\mathbf{B}|$ represent the magnitudes of vectors **A** and **B** and θ is the angle between vectors **A** and **B**.

• The directional derivative $f_a(x_1,y_1)$ will have its greatest value when its unit vector **a** is pointing in the same direction as the gradient of $f(x_1,y_1)$. Therefore, the directional derivative will be greatest when the *angle* θ between it and the gradient is zero.

In other words, the slope of the directional derivative $f_a(x_1,y_1)$ is greatest when **a** is *parallel* to (grad f). Therefore, writing the directional derivative in terms of the dot product gives:

$$f_a(x_1,y_1) = ((\partial f/\partial x)\mathbf{i} + (\partial f/\partial y)\mathbf{j}) \bullet \mathbf{a}$$
$$= |((\partial f/\partial x)\mathbf{i} + (\partial f/\partial y)\mathbf{j})||a| \cos \theta = |((\partial f/\partial x)\mathbf{i} + (\partial f/\partial y)\mathbf{j})| \cos \theta$$

θ will be zero when $f_a(x_1,y_1)$ has its greatest value:

$$f_a(x_1,y_1) = |((\partial f/\partial x)\mathbf{i} + (\partial f/\partial y)\mathbf{j})| \cos \theta = |((\partial f/\partial x)\mathbf{i} + (\partial f/\partial y)\mathbf{j})|$$

Therefore, $f_a(x_1,y_1) = |\text{grad } f(x_1,y_1)|$ when they both have the same magnitude (at the maximum value) and **a** is pointing in the same direction as grad $f(x_1,y_1)$.

The greatest *slope* is equivalent to the magnitude $|\text{grad } f| = \sqrt{f_x^2 + f_y^2}$ and occurs when (grad f) $\bullet$ **a** $= |\text{grad } f|$.

• The *directional derivative* will have a zero rate of change when the angle θ between it and the gradient is 90 degrees, where $\theta = \pi/2$ and **a** is pointing perpendicular to (grad f). Therefore, when $\theta = \pi/2$:

$$\text{grad } f(x_1,y_1) \bullet \mathbf{a} = |\text{grad } f(x_1,y_1)| \cos \pi/2$$
$$= |((\partial f/\partial x)\mathbf{i} + (\partial f/\partial y)\mathbf{j})| \cos \pi/2 = 0$$

• The directional derivative will have its most negative rate when the angle θ between it and the gradient is 180 degrees, where $\theta = \pi$ and **a** is pointing in the opposite direction of (grad f). Therefore, when $\theta = \pi$:

$$\text{grad } f(x_1,y_1) \bullet \mathbf{a} = \left| \text{grad } f(x_1,y_1) \right| \cos \pi$$
$$= \left| ((\partial f/\partial x)\mathbf{i} + (\partial f/\partial y)\mathbf{j}) \right| \cos \pi = -\left| \text{grad } f(x_1,y_1) \right|$$

• When f(x,y) is a *linear function*, the gradient is a constant vector because the terms $\partial/\partial x$ and $\partial/\partial y$ will yield constants. Conversely, when f(x,y) is a *non-linear function*, the gradient is a non-constant or varying vector.

6.6 Minima, Maxima, and Optimization

• This section introduces minima and maxima problems for functions having more than one variable and is an extension of Section 2.27 for one-variable functions. This section includes the first and second derivatives for surfaces, finding minima and maxima points, and the concept of constrained optimization.

• Evaluating whether a function has *minimum* and *maximum* points is common when experiments or evaluations are conducted in science, business, engineering, etc. Data is gathered, relationships are developed, and graphs are constructed in order to assist in the understanding of the data and to predict future patterns and events. Information depicted in the graphs such as where the graph is rising or falling, convex or concave, and where the high and low points are (which correspond to the maximum and minimum values) are all crucial to the evaluation of the data.

• The graph of a function has a minimum or maximum point where the *slope is zero* and, therefore, the *derivative is zero*. In the region of a graph of a function where the graph is horizontal, the first derivative of the function is equal to zero. A point where the graph of a function is horizontal may represent a *minimum* or *maximum* point. A minimum or maximum on a graph may be the minimum or maximum of the function, *global extrema*, or there may be many "local" minimum or maximum points called *local extrema*. There are also examples where a graph will not have a minimum or maximum, such as if the graph is a straight horizontal or vertical line or plane.

• For a function with a single variable $y = f(x)$, a minimum or maximum point occurs where $df/dx = 0$. For a function with more than one variable $z = f(x,y)$, a minimum or maximum point occurs where $(\partial f/\partial x) = 0$, $(\partial f/\partial y) = 0$, the appropriate partial derivatives of the independent variables. Where the graph of a multivariable function is level, the *partial derivatives are all zero.*

• In a *one-variable function* (discussed in Section 2.27), the *sign of the derivative* of the function indicates the slope of the graph of the function at the point where the derivative is taken. For $y = f(x)$, $f'(x) < 0$ where the graph of f is decreasing, $f'(x) > 0$ where the graph of f is increasing, and $f'(x) = 0$ where the graph of f is horizontal. The sign of $f'(x)$ changes from positive to negative or negative to positive as the maximum or minimum is crossed. There are functions that don't possess minimum or maximum points, such as where an inflection point exists.

• In the *one-variable* case, taking the *second derivative* of a function is used to determined whether the graph of that function is at a minimum and, therefore, *concave up*, or at a maximum and, therefore, *concave down*. For $f(x)$ at point P, where $f'(P)$ exists and $f'(P) = 0$, then if $f''(P) > 0$, the graph of the function is concave up at P and has a minimum at P. Conversely, if $f'(P) = 0$ and if $f''(P) < 0$, the graph of the function is concave down at P and has a maximum at P. See Section 2.27 for a complete discussion of minima and maxima for single variable functions.

• For a function $z = f(x,y)$ with *two independent variables*, a maximum on the graph of that function exists at a point (x_1,y_1) if $f(x,y) \leq f(x_1,y_1)$ for all values of x and y near (x_1,y_1). Conversely, a minimum exists where $f(x,y) \geq f(x_1,y_1)$ for all values of x and y near (x_1,y_1). To summarize, *global and local extrema* occur for $f(x,y)$ according to the following:

> *Global maximum* exists at (x_1,y_1) if $f(x,y) \leq f(x_1,y_1)$ for all (x,y);
>
> *Global minimum* exists at (x_1,y_1) if $f(x,y) \geq f(x_1,y_1)$ for all (x,y);
>
> *Local maximum* exists at (x_1,y_1) if $f(x,y) \leq f(x_1,y_1)$ for (x,y) near (x_1,y_1); and
>
> *Local minimum* exists at (x_1,y_1) if $f(x,y) \geq f(x_1,y_1)$ for (x,y) near (x_1,y_1).

- The following properties of f(x,y) can be compared with f(x):

(a.) The extrema points occur at $(\partial f/\partial x) = 0$ and $(\partial f/\partial y) = 0$ rather than df/dx = 0.

(b.) A *tangent plane* exists where derivatives are zero rather than a tangent line.

(c.) A boundary *curve* encompasses the region of interest rather than two endpoints.

(d.) Partial derivatives $(\partial^2 f/\partial x^2)$, $(\partial^2 f/\partial xy)$, and $(\partial^2 f/\partial y^2)$ are used to determine whether the extrema is a minimum or maximum or a saddle point, rather than using ordinary derivatives d^2f/dx^2, d^2f/dxy, and d^2f/dy^2.

- In a closed and bounded region, a continuous function f(x,y) will generally have a global minimum and a global maximum. A *closed region* contains a boundary and if a region is *bounded*, then it does not go to infinity in any direction. If a region is not closed and bounded or f(x,y) is not a continuous function, there may or may not be a global minimum or global maximum present.

- Local extrema generally occur at *critical points* where the derivative is zero or undefined. For a minimum or maximum to exist for z = f(x,y) at (x_1,y_1), it is necessary that $(\partial f/\partial x) = 0$ and $(\partial f/\partial y) = 0$ (for three variables, include $(\partial f/\partial z) = 0$). Note that this condition is not sufficient to assure that a minimum or maximum exists.

- If $f(x_1,y_1)$ is a minimum or maximum point, then the *gradient vector* at that point will be zero. The slope in every direction will be zero. Therefore, (grad $f(x_1,y_1)$) equals zero or is undefined at a minimum or maximum point. *Critical points* occur where the gradient is either zero or undefined.

- To *determine if maxima or minima exist* for z = f(x,y) the following steps can be taken:

(a.) Calculate $(\partial f/\partial x)$, $(\partial f/\partial y)$, $(\partial^2 f/\partial x^2)$, $(\partial^2 f/\partial y^2)$, and $(\partial^2 f/\partial xy)$.

(b.) Solve $(\partial f/\partial x) = 0$ and $(\partial f/\partial y) = 0$ simultaneously for the critical values of x and y that satisfy these equations. (In this case there are two equations and two unknowns.)

(c.) Determine the value given by:

$D = (\partial^2 f/\partial x^2)(\partial^2 f/\partial y^2) - (\partial^2 f/\partial xy)^2$ at point (x_1, y_1).

(d.) Evaluate the following criteria at point (x_1, y_1) for $z = f(x,y)$:

Minimum if $D > 0$ and $(\partial^2 f/\partial x^2) > 0$ or $(\partial^2 f/\partial y^2) > 0$;

Maximum if $D > 0$ and $(\partial^2 f/\partial x^2) < 0$ or $(\partial^2 f/\partial y^2) < 0$;

No minimum or maximum if $D < 0$ (saddle point);

This test fails if $D = 0$.

• **Example:** Does a minimum or maximum exist for $z = x^2 + y^2$?

Calculate:

$(\partial f/\partial x) = 2x$
$(\partial f/\partial y) = 2y$
$(\partial^2 f/\partial x^2) = 2$
$(\partial^2 f/\partial y^2) = 2$
$(\partial^2 f/\partial xy) = 0$

Solve:

$2x = 0 \rightarrow x = 0$
$2y = 0 \rightarrow y = 0$

Determine:

$D = (\partial^2 f/\partial x^2)(\partial^2 f/\partial y^2) - (\partial^2 f/\partial xy)^2 = (2)(2) - 0 = 4$

Because $4 = D > 0$ and $(\partial^2 f/\partial x^2) = 2 > 0$, then a minimum exists at $(x = 0, y = 0, z = 0)$.

• **Example:** Does a minimum or maximum exist for

$z = x^2 - y^2$?

Calculate:

$(\partial f/\partial x) = 2x$
$(\partial f/\partial y) = -2y$
$(\partial^2 f/\partial x^2) = 2$
$(\partial^2 f/\partial y^2) = -2$
$(\partial^2 f/\partial xy) = 0$

Solve:

$$2x = 0 \rightarrow x = 0$$
$$2y = 0 \rightarrow y = 0$$

Determine:

$$D = (\partial^2 f/\partial x^2)(\partial^2 f/\partial y^2) - (\partial^2 f/\partial xy)^2 = -4 < 0$$

Because $-4 = D < 0$, then no minimum or maximum exists and this is a *saddle point* that corresponds to an inflection point for a single variable function. At a saddle point there are values of x and y such that $f(x_1,y_1) > f(x,y)$ and also $f(x_1,y_1) < f(x,y)$.

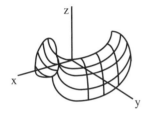

- The graph of a *quadratic function* $f(x,y) = ax^2 + bxy + cy^2$ can be analyzed for minima, maxima, and saddle points by using a technique that involves completing the square of $ax^2 + bxy + cy^2$ and results in the following being true at point (0,0):

   Minimum exists at (0,0) when $a > 0$ and $(4ac - b^2) > 0$;
   Maximum exists at (0,0) when $a < 0$ and $(4ac - b^2) > 0$;
   A saddle point exists at (0,0) when $(4ac - b^2) < 0$.

Note that for a point at (x_1,y_1) rather than (0,0), the quadratic function will have the form:

$$f(x,y) = a(x - x_1)^2 + b(x - x_1)(y - y_1) + c(y - y_1)^2 + d$$

and the graph will have the same shape as it would at point (0,0) except it will be located at point (x_1,y_1) and shifted the value of d in the vertical direction.

Constrained Optimization

- The following paragraphs provide a brief introduction to constrained optimization. For a complete discussion of constrained optimization, a more advanced mathematical analysis book should be consulted.

• When a system or graph is evaluated using optimization techniques (minimization and maximization), there is often more than one function involved in describing the system or graph. When minimizing or maximizing, it can be beneficial to hold one function constant or *constrained* while considering the other function. Finding local minima or maxima for the two functions f(x,y) and g(x,y) involves finding the partial derivatives $(\partial f/\partial x)$, $(\partial f/\partial y)$, $(\partial g/\partial x)$, and $(\partial g/\partial y)$. When g(x,y) is *constrained* or held constant such that g(x,y) = C, the extrema of f(x,y) has the following properties:

(a.) f(x,y) has a global minimum at some point (x_1,y_1) when
     $f(x,y) \geq f(x_1,y_1)$ for all values of x and y.

(b.) f(x,y) has a global maximum at some point (x_1,y_1) when
     $f(x,y) \leq f(x_1,y_1)$ for all values of x and y.

(c.) f(x,y) has a local minimum at some point (x_1,y_1) when
     $f(x,y) \geq f(x_1,y_1)$ for values of x and y near (x_1,y_1).

(d.) f(x,y) has a local maximum at some point (x_1,y_1) when
     $f(x,y) \leq f(x_1,y_1)$ for values of x and y near (x_1,y_1).

• To evaluate a *constrained optimization* problem, the local extrema of one function f(x,y) can be found while the other function g(x,y) is constrained such that g(x,y) = C. The extrema found using such a constraint may not be the same extrema present if no constraint was present. Also, determining whether the extrema is a minimum or maximum can be observed by graphing the functions.

• Consider the graph of two functions f(x,y) and g(x,y) that are related to each other by a scalar quantity called λ (lambda), which is known as the *Lagrange multiplier*. When f(x,y) is at a minimum or maximum point with the constraint g(x,y) = C, the gradient of f is parallel to the gradient of g. At a minimum or maximum point, (grad f) and (grad g) are related to each other by the multiplier λ, such that for g = C, the following is true:

grad f = λ grad g

$(\partial f/\partial x) = \lambda(\partial g/\partial x)$

$(\partial f/\partial y) = \lambda(\partial g/\partial y)$

• To find extrema for f, the three equations g = C, $(\partial f/\partial x) = \lambda(\partial g/\partial x)$, and $(\partial f/\partial y) = \lambda(\partial g/\partial y)$ can be solved for the three unknown values, x, y, and λ.

- For a function $f(x,y,z)$ with two constraints, $g(x,y,z) = C_1$ and $h(x,y,z) = C_2$, there are two multipliers λ_1 and λ_2. To minimize or maximize f, the following equations can be solved for x, y, z, λ_1, and λ_1:

$$(\partial f/\partial x) = \lambda_1(\partial g/\partial x) + \lambda_2(\partial h/\partial x)$$
$$(\partial f/\partial y) = \lambda_1(\partial g/\partial y) + \lambda_2(\partial h/\partial y)$$
$$(\partial f/\partial z) = \lambda_1(\partial g/\partial z) + \lambda_2(\partial h/\partial z)$$
$$g = C_1 \text{ and } h = C_2$$

- Optimization problems are sometimes written in terms of a *Lagrangian function L*: $L(x,y,\lambda) = f(x,y) - \lambda(g(x,y) - C)$.

The solution for a constrained optimization problem involving L is found using:

$$(\partial L/\partial x) = (\partial f/\partial x) - \lambda(\partial g/\partial x) = 0$$
$$(\partial L/\partial y) = (\partial f/\partial y) - \lambda(\partial g/\partial y) = 0$$
$$(\partial L/\partial \lambda) = C - g = 0$$

At a critical point (x_1,y_1,λ_1) of $f(x,y)$ where $g(x,y) = C$ and λ_1 is the corresponding Lagrange multiplier: grad $L(x_1,y_1,\lambda_1) = 0$.

- There are constraints involving *inequalities* such as $g \leq C$ or $g \geq C$, where the multiplier λ must satisfy the same inequalities, such that $\lambda \leq C$ or $\lambda \geq C$. For example, if the constraint is $g \leq C$, then the extrema can be inside or on the constraint curve.

Chapter

7

Vector Calculus

This chapter is designed to provide definitions, formulas, and brief explanations that are important in vector calculus, and also provide a context for how the topics described fit into the subject of calculus.

7.1 Summary of Scalars, Vectors, the Directional Derivative, and the Gradient

• This section provides a brief summary of scalars, vectors, the directional derivative, and the gradient. See Chapter 5 for more information on vectors and Chapter 6 for more information on the directional derivative and the gradient.

Scalars and Vectors

• *Scalar functions* are functions whose values are *scalars*. Similarly, *vector functions* are functions whose values are *vectors*. A scalar function in three dimensions $f = f(x,y,z)$ is defined at some point (x,y,z) by a value, whereas a vector function in three dimensions $\mathbf{v} = \mathbf{v}(x,y,z)$ has three components such that $\mathbf{v} = [v_1(x,y,z), v_2(x,y,z), v_3(x,y,z)]$. A vector function has an input point (x,y,z) and an output that has a three-dimensional vector function that represents a field of vectors with one at each point in the field.

• Both scalar and vector functions are used in applications where the *domain* of a function is a curve in space, a surface in space, or some region in space on a curve or surface. A scalar function defines a *scalar field* or a region on a curve or surface. Examples include temperature fields and pressure fields. A vector function defines a *vector field* and has a vector at each point in two- or three-dimensional space in a region, curve, surface, or volume. Examples include velocity fields, force fields, and gravitational fields.

• Vector and scalar functions sometimes depend on time t or other parameters. An example of a scalar function $f(x,y,z)$ is the *distance* of a point $P_0 = (x_0,y_0,z_0)$ to another point $P = (x,y,z)$.

The domain is all the space and $f(x,y,z)$ defines the scalar field in space:

$$\text{distance} = f(x,y,z) = \sqrt{(x - x_0)^2 + (y - y_0)^2 + (z - z_0)^2}$$

This formula is given in Cartesian coordinates. However, the distance would be the same if represented in another coordinate system. See Section 7.2 for a discussion of vectors fields.

Directional Derivative and Gradient

• A surface can have slopes in all directions, not just along the axes. The *directional derivative* represents the slope of a tangent line to a surface at a point in any specified direction. As discussed in the Chapter 6, $(\partial z/\partial x)$ and $(\partial z/\partial y)$ represent the rate of change of a surface $z = f(x,y)$ in the directions of the X-axis and Y-axis, respectively.

• A small change ds along a surface $z = f(x,y)$ in a given direction can be represented using a unit vector **a** that is pointed in a designated direction and has components a_1 and a_2. Components a_1 and a_2 can correspond to **i** and **j** unit vectors and **a** can be represented by: $\mathbf{a} = a_1\mathbf{i} + a_2\mathbf{j}$

The *directional derivative* df/ds in the direction of vector **a** can be written:

$$dz/ds = (\partial f/\partial x)a_1 + (\partial f/\partial y)a_2$$

Remember that **i**, **j**, and **k** are *unit vectors* that point parallel to the axes of a coordinate system. (See Section 5.1.)

• *Notation for the directional derivative* includes df/ds, dz/ds, $D_a f$ and $f_a(x_1,y_1)$. For example, using the notation $f_a(x_1,y_1)$, the directional derivative in the direction of vector **a** is written:

$$f_a(x_1,y_1) = f_x(x_1,y_1)a_1 + f_y(x_1,y_1)a_2$$

• The *gradient* is a vector quantity and describes the change in a function near a point. The *gradient characterizes maximum increase and points in the direction of maximum increase of f at a selected point.* The gradient of $f(x,y)$ is written:

$$\text{grad } f = (\partial f/\partial x)\mathbf{i} + (\partial f/\partial y)\mathbf{j}$$

For $f(x,y,z)$, the gradient becomes:

$$\text{grad } f = (\partial f/\partial x)\mathbf{i} + (\partial f/\partial y)\mathbf{j} + (\partial f/\partial z)\mathbf{k}$$

The components of the vector (grad f) are:

$$(\partial f/\partial x)\mathbf{i}, \; (\partial f/\partial y)\mathbf{j}, \text{ and } (\partial f/\partial z)\mathbf{k}$$

• The *gradient* of a *scalar function* $f(x,y,z)$ is a *vector function*. The gradient of $f(x,y,z)$ at point P denotes the *direction of maximum increase*, providing grad $f \neq 0$ and f is differentiable.

• The *gradient vector* of $z = f(x,y)$ at point (x_1,y_1) points in the direction of where the greatest change in $f(x,y)$ occurs (providing f is differentiable at (x_1,y_1)). The *gradient lies on the surface of f(x,y) in the x-y plane and points in the direction that the surface is rising or increasing the most.* In a *contour diagram*, the gradient vector points perpendicular to the contour lines in the direction of greatest increase in height, which is where the contour lines are closest together.

• The *magnitude or length of the gradient vector* given by $|\text{grad f}|$ is equal to the rate of change in the direction that it is pointing. On a contour diagram, the gradient vector has a magnitude corresponding to the degree (or grade) of the *slope*. Because the slope is greater when the contour lines are closer together, the magnitude of the gradient vector is greater for contours that are closer together. Conversely, because the slope is less when the contour lines are more separated, the magnitude of the gradient vector is smaller for contours that are farther apart.

• *Notation for grad* is ∇, whic\h is also called *"del"* and is a vector that is an *operator* because its components are operations rather than numbers.

$$\nabla = (\partial/\partial x)\mathbf{i} + (\partial/\partial y)\mathbf{j} + (\partial/\partial z)\mathbf{k}$$

Therefore, grad $f = \nabla f = (\partial f/\partial x)\mathbf{i} + (\partial f/\partial y)\mathbf{j} + (\partial f/\partial z)\mathbf{k}$.

Notation, for the gradient of f includes:

grad $f(x,y,z) = (\partial f/\partial x)\mathbf{i} + (\partial f/\partial y)\mathbf{j} + (\partial f/\partial z)\mathbf{k}$

grad $f(x_1,y_1,z_1) = f_x(x_1,y_1,z_1)\mathbf{i} + f_y(x_1,y_1,z_1)\mathbf{j} + f_z(x_1,y_1,z_1)\mathbf{k}$

• **Example:** If $f = 3x + 2yz - 6y^2$, what is grad f?

grad $f = \nabla f = (3)\mathbf{i} + (2z - 12y)\mathbf{j} + (2y)\mathbf{k}$

• The *dot product* of the *gradient vector* at point (x_1,y_1) with the unit vector **a** is equal to the *directional derivative* $f_a(x_1,y_1)$ pointing in the direction of **a** at point (x_1,y_1):

grad $f(x_1,y_1) \bullet \mathbf{a} = f_a(x_1,y_1)$

$= ((\partial f/\partial x)\mathbf{i} + (\partial f/\partial y)\mathbf{j}) \bullet (a_1\mathbf{i} + a_2\mathbf{j}) = (\partial f/\partial x)a_1 + (\partial f/\partial y)a_2$

$= |\text{grad } f(x_1,y_1)|\cos \theta = |((\partial f/\partial x)\mathbf{i} + (\partial f/\partial y)\mathbf{j})|\cos \theta$

where $\mathbf{a} = a_1\mathbf{i} + a_2\mathbf{j}$.

Remember that the dot product of two vectors is:

$$\mathbf{A} \bullet \mathbf{B} = |\mathbf{A}||\mathbf{B}|\cos \theta$$

where $|\mathbf{A}|$ and $|\mathbf{B}|$ represent the magnitudes of vectors **A** and **B** and θ is the angle between vectors **A** and **B**.

• The directional derivative $f_a(x_1,y_1)$ will have its greatest value when its unit vector **a** is pointing in the same direction as the gradient of $f(x_1,y_1)$. Therefore, the directional derivative will be greatest when the *angle* θ between it and the gradient is zero.

7.2 Vector Fields and Field Lines

• This section includes definitions of vector fields and examples of vector fields including horizontal, radial, rotational, gradient, force, velocity and flow, and also the definition of field lines.

• *Vector functions or fields* have an input as point (x,y) or point (x,y,z) and an output as a two- or three-dimensional vector function **F**(x,y) or **F**(x,y,z) that represents a field of vectors with one at each point in the field. A vector function defines a *vector field*, which has a vector at each point in two- or three-dimensional space in a region, curve, surface, or volume. Examples include velocity fields, force fields, and gravitational fields. The following are examples of geometric configurations of vector fields:

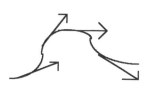

tangent vector field
on curve

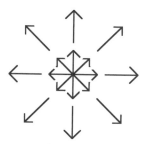

vector field of rotating body

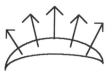

normal vector field
on surface

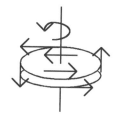

radial vector field

• A *vector field* is a function that possesses a vector at each point in a two-dimensional plane or three-dimensional space. In a vector function or vector field, the value of the field at any point is the vector denoting *magnitude* and *direction*. In two-dimensions, a vector field is a vector function **F**(x,y) whose value at any point (x,y) is a two-dimensional vector.

Similarly, in three dimensions a vector field is a vector function $F(x,y,z)$ whose value at any point (x,y,z) is a three-dimensional vector. The values of $F(x,y)$ and $F(x,y,z)$ are two- and three-dimensional vectors.

• A *point* in a *vector field* can be represented by its *position vector* **R**. Therefore, a vector field is sometimes represented by $F(R)$. Also, a vector field can be represented by the function **F** describing the field. (See Section 5.1 for a definition of the position vector.)

• In two dimensions $F(x,y)$ has two components and in three dimensions $F(x,y,z)$ has three components:

$$F(x,y) = F_1(x,y)\mathbf{i} + F_2(x,y)\mathbf{j}$$
$$F(x,y,z) = F_1(x,y,z)\mathbf{i} + F_2(x,y,z)\mathbf{j} + F_3(x,y,z)\mathbf{k}$$

• The components of a vector do not vary. However, the components in a vector field are variable.

• The following are examples of vector fields.

(a.) *Horizontal field*: $F(x,y) = x\mathbf{i}$

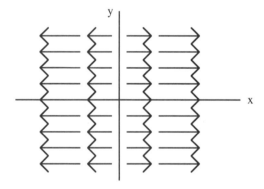

Vector $x\mathbf{i}$ is parallel to the X-axis and points in the positive x-direction when x is positive and points in the negative x-direction when x is negative. Because **F** does not depend on y, the vectors along the Y-axis direction are the same length. In general, longer vectors have a larger magnitude.

(b.) *Radial field*: $\mathbf{R}(x,y) = x\mathbf{i} + y\mathbf{j}$

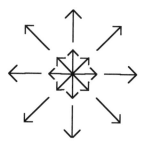

The *position vector* $\mathbf{R}$ at point (x,y) describes a radial field with components $R_1 = x$ and $R_2 = y$. The length of the vectors are longer further from the origin and are given by:

$$|\mathbf{R}| = (x^2 + y^2)^{1/2}$$

(c.) *Rotation field*: $\mathbf{S}(x,y) = -y\mathbf{i} + x\mathbf{j}$

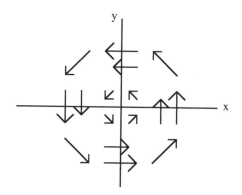

This is a rotation or spin field with components $S_1 = -y$ and $S_2 = x$. The length, $|\mathbf{S}| = ((-y)^2 + x^2)^{1/2}$, is the distance from (x,y) to the origin. Vectors at each fixed distance from the origin have the same magnitude, and the magnitude increases further from the origin. At each point (x,y) vector $\mathbf{S}$ is perpendicular to the *position vector* $\mathbf{R} = x\mathbf{i} + y\mathbf{j}$. Because $\mathbf{S}$ is perpendicular to $\mathbf{R}$, $\mathbf{S} \cdot \mathbf{R} = -yx + xy = 0$.

(d.) *Gradient field*: $\mathbf{F} = \text{grad } f = (\partial f/\partial x)\mathbf{i} + (\partial f/\partial y)\mathbf{j}$

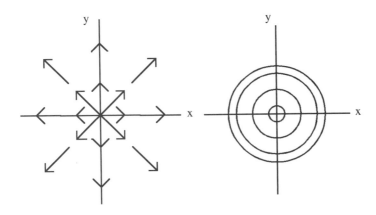

This represents the gradient field of scalar function f with components $F_1 = \partial f/\partial x$ and $F_2 = \partial f/\partial y$ and its *contour diagram*. This vector field is the gradient of function f(x,y) where at each point (x,y) the vector (grad f) points in the direction of maximum rate of increase of f(x,y). By definition of a gradient, the vectors in a gradient field are perpendicular to the level curves (contours) f(x,y) = c and pointing in the direction of increasing f. The length (or magnitude) | grad f | represents the rate of change of f in the direction of increasing f. The rate of change is larger when the contours are closer together.

In a gradient vector field (grad f) the scalar function f is called a *potential function* of the vector field. Gradient fields are also called *conservative* because in a gradient field, energy is conserved and, therefore, no energy is gained or lost when displacement of an object or charge occurs from an initial point P to another point in the field and back to the initial point P.

Note that a *radial vector* field **R** is also a gradient field. *Gradient fields* also include vector fields in the form $\mathbf{R}/r^n$. The vector fields **S** and **S**/r are *rotation or spin fields* and not gradient fields.

(e.) *Force fields*: Force fields include *gravitational fields* such as Earth's gravitational force on all other masses. The direction of the Earth's gravitational field is toward its center, and the magnitude decreases further from Earth. Note that a gravitational force field is *conservative*.

(f.) *Velocity field and flow field*: In a *velocity vector field* each vector represents the velocity of the flow at that point. The flow is fastest where the velocity vectors are longest, which generally occurs in the center of a flow stream. For example, in a fluid moving steadily inside a pipe, the velocity can be different at different points. A velocity field can be horizontal, rotational, radial, etc.

The velocity vector **V** provides the speed and direction of flow at each point in the field. In three-dimensional flow the vector field **V**(x,y,z) has three components V_1, V_2, V_3. The velocity field is $V_1\mathbf{i} + V_2\mathbf{j} + V_3\mathbf{k}$ and speed or length is:

$$|\mathbf{V}| = \sqrt{(V_1)^2 + (V_2)^2 + (V_3)^2}$$

A *flow field* has density ρ multiplied by the velocity **V**, or ρ**V**. In a flow field ρ**V**, **V** represents the rate of movement and ρ**V** is the rate of movement of mass. A greater density yields a greater $|\rho\mathbf{V}|$ of mass transport.

Field Lines

• *Field lines* are the curves or lines that are tangent to the vectors in a vector field. For example, in a *rotation field* the field lines are circles and in a *gravity field* or a *radial field* the field lines are rays extending from the origin. Field lines are also referred to as *integral curves, streamlines,* and *flow lines.* Note that the lengths of the vectors in a vector field are not represented by the field lines.

• In a *gradient field* $\mathbf{F} = (\partial f/\partial x)\mathbf{i} + (\partial f/\partial y)\mathbf{j}$, the vector field is tangent to the *field lines,* and the *level curves* (*contour lines*), also called *equipotentials*, are perpendicular to the field lines. A gradient field **F**(x,y) has a *potential* f(x,y) and it has level curves that connect points that have equal potential and are called *equalpotentials.*

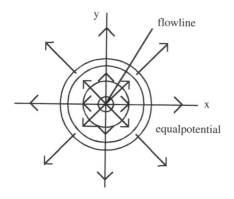

• In a *velocity vector field*, each fluid particle moves along a field line (or stream line). The flow in a velocity field is represented by the family of all of its flow field lines. If a particle is moving in a velocity field along the surface of water, the velocity of the particle at time t is equal to the velocity of the fluid at the particle's position at time t. The flow line can be found using the position vector $\mathbf{R}(t)$ of the particle at time t, where $d\mathbf{R}/dt$ is the velocity of a fluid particle at time t:

$$d\mathbf{R}(t)/dt = \mathbf{F}(\mathbf{R}(t))$$

where $\mathbf{F} = F_1\mathbf{i} + F_2\mathbf{j}$, $\mathbf{R}(t) = x(t)\mathbf{i} + y(t)\mathbf{j}$, and $dx/dt = F_1$ and $dy/dt = F_2$.

Note that $x(t)$ and $y(t)$ (or equivalently $\mathbf{R}(t)$) describe the *path of motion*. Solving $dx/dt = F_1$ and $dy/dt = F_2$ for $x(t)$ and $y(t)$ provides a *parameterization* of the flow line or path of motion of a particle and the flow line at a specified point.

• *Flow lines* in a velocity field can be approximated using *Euler's method* of solving differential equations. Using flow lines $\mathbf{R}(t) = x(t)\mathbf{i} + y(t)\mathbf{j}$ of vector field $\mathbf{F}(x,y)$, where $d\mathbf{R}(t)/dt = \mathbf{F}(\mathbf{R}(t))$ is the differential equation, then:

$$\mathbf{R}(t + \Delta t) \approx \mathbf{R}(t) + (\Delta t)d\mathbf{R}/dt = \mathbf{R}(t) + (\Delta t)\mathbf{F}(\mathbf{R}(t)) \text{ for } \Delta t \text{ near } 0$$

To approximate the flow line, begin at point $\mathbf{R}_0 = \mathbf{R}(0)$ and estimate the next position $\mathbf{R}_1$ of a particle at $t = \Delta t$:

$$\mathbf{R}_1 = \mathbf{R}(\Delta t) \approx \mathbf{R}(0) + (\Delta t)\mathbf{F}(\mathbf{R}(0)) = \mathbf{R}_0 + (\Delta t)\mathbf{F}(\mathbf{R}_0)$$

At $\mathbf{R}_{n+1}$ for subsequent positions $\mathbf{R}_0$, $\mathbf{R}_1$, $\mathbf{R}_2$, etc., that represent the path use:

$$\mathbf{R}_{n+1} = \mathbf{R}_n + (\Delta t)\mathbf{F}(\mathbf{R}_n) = \mathbf{R}_n + (\Delta t)\mathbf{F}(x_n, y_n)$$

where $\mathbf{R}_n = x_n\mathbf{i} + y_n\mathbf{j}$ and $d\mathbf{R}/dt = \mathbf{F}$.

The vectors $\mathbf{R}_n$ establish the path or flow line.

7.3 Line Integrals and Conservative Vector Fields

• This section includes the definition of a line integral, the line integral of a vector field along a curve, independence of the path of a line integral, and conservative gradient fields and the line integral.

• A *line integral* is an integral along a *curve* and is a generalization of a *definite integral*. Remember the definite integral $_a\!\int^b$ f(x) dx, where integration occurs along the X-axis from point a to point b of the integrand f, which is a function existing at each point between a and b.

Similarly, for a line integral $\int_C \mathbf{F} \bullet d\mathbf{R}$, integration occurs along a curve C in a plane or in three-dimensional space where the integrand is a function existing at each point along the curve. The curve is called the *path of integration*. The *orientation* of a curve is the direction of motion or travel along the curve.

• The line integral represents the *work along a curve* and is used in *Green's Theorem* (Section 7.4) and *Stoke's Theorem* (Section 7.8), which connect line integrals to surface integrals. *Applications* of the *line integral* include the work between two points, the work during a change in kinetic energy, and work done by gravity on an object in motion. The line integral around a closed curve also represents *circulation*, which is a measure of the extent to which the vector field points around the closed curve.

• To *develop the line integral*, consider a curve that is smooth and continuous and is oriented so that it begins at point a and ends at point b. This curve can be represented using the position vector:

$$\mathbf{R}(t) = x(t)\mathbf{i} + y(t)\mathbf{j} + z(t)\mathbf{k} \qquad (a \leq t \leq b)$$

where $\mathbf{R}(t)$ is smooth and continuous and $d\mathbf{R}/dt \neq 0$. If points a and b coincide, then the curve is a *closed curve*.

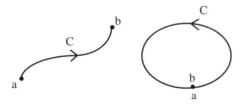

Similar to a definite integral, a line integral can be thought of as consisting of a sum of infinitely many tiny smooth curves between points a and b on curve C. For a vector field $\mathbf{F}$ and curve C, C can be segmented in small sections that are approximately straight and where $\mathbf{F}$ is approximately constant, such that each section can be represented by displacement vector $\Delta\mathbf{R}_i = \mathbf{R}_{i+1} - \mathbf{R}_i$. At each point $\mathbf{R}_i$, the dot product $(\mathbf{F}(\mathbf{R}_i) \bullet \Delta\mathbf{R}_i)$ compares $\Delta\mathbf{R}_i$ with the value of vector field $\mathbf{F}(\mathbf{R}_i)$. The sum of all sections of C is:

$$\sum \mathbf{F}(\mathbf{R}_i) \bullet \Delta \mathbf{R}_i$$

The limit as $|\Delta \mathbf{R}_i| \to 0$ results in the line integral:

$$\lim_{|\Delta \mathbf{R}i| \to 0} \sum \mathbf{F}(\mathbf{R}_i) \bullet \Delta \mathbf{R}_i = \int_C \mathbf{F} \bullet d\mathbf{R}$$

which is the *line integral*.

Therefore, the *line integral of vector function* $\mathbf{F}(\mathbf{R})$ over curve C, is defined as:

$$\int_C \mathbf{F}(\mathbf{R}) \bullet d\mathbf{R} = \int_C (F_1 dx + F_2 dy + F_3 dz)$$
$$= {}_a\!\int^b \mathbf{F}(\mathbf{R}(t)) \bullet (d\mathbf{R}/dt)dt$$
$$= {}_a\!\int^b (F_1(dx/dt) + F_2(dy/dt) + F_3(dz/dt))dt$$

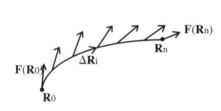

- *Notation for the line integral* over curve C is: $\int_C$. If curve C is a *closed curve*, the integral symbol is often written: $\oint_C$.

- The definition of the *line integral* depends on $\mathbf{F}$ being a continuous open set containing curve C, which is a smooth continuous curve that can be parameterized. *Parameterization* of a curve proceeds from the beginning point to the ending point without retracing.

- The *line integral of vector field* $\mathbf{F}$ along curve C indicates the extent that C is going with or against $\mathbf{F}$. The line integral, therefore, depends on the values of the vector field along curve C. Because the line integral of $\mathbf{F}$ sums dot products with $d\mathbf{R}$ (or $\Delta \mathbf{R}_i$) along a curve, then the following are true: (a.) If $\mathbf{F}$ is generally pointing in the same direction as C at all points along C, then the result is positive; (b.) if $\mathbf{F}$ is generally pointing in the opposite direction, the result is negative; and (c.) if $\mathbf{F}$ is perpendicular to C at all points, then the result is zero.

- *Properties of line integrals* include the following:

$$\int_C k\mathbf{F} \bullet d\mathbf{R} = k\int_C \mathbf{F} \bullet d\mathbf{R} \text{ , where k is a constant.}$$

$$\int_C (\mathbf{F} + \mathbf{G}) \bullet d\mathbf{R} = \int_C \mathbf{F} \bullet d\mathbf{R} + \int_C \mathbf{G} \bullet d\mathbf{R}$$

$$\int_{-C} \mathbf{F} \bullet d\mathbf{R} = -\int_C \mathbf{F} \bullet d\mathbf{R}$$

where integrating $\mathbf{F}$ along C in the opposite direction is the negative of the line integral along C.

$$\int_C \mathbf{F} \bullet d\mathbf{R} = \int_{C_1} \mathbf{F} \bullet d\mathbf{R} + \int_{C_2} \mathbf{F} \bullet d\mathbf{R}$$

where C_1 and C_2 combine to C.

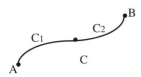

- An example of an important line integral is *work along a curve*. The line integral is:

$$\int_C \mathbf{F} \bullet d\mathbf{R} = {_a}\!\int^b \mathbf{F}(\mathbf{R}(t)) \bullet (d\mathbf{R}/dt)dt$$

where t is the arc length of C and the tangential component of $\mathbf{F}$. The work is done by force $\mathbf{F}$ in a displacement along C. Work is done in the direction of movement.

If *displacement* occurs along a *straight line*, the *work* done by a constant force $\mathbf{F}$ is: Work $= \mathbf{F} \bullet \mathbf{d} = |\mathbf{F}||\mathbf{d}|\cos\theta$

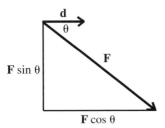

If displacement occurs along a *curve*, the work done by a variable force **F** is the sum of work done in displacement along small curves (or segments of curve C). If a force at a point with position vector **R** given by **F(R)** is acting on an object moving along curve C, then work done by force **F(R)** over a small distance Δ**R** is **F(R)** • Δ**R** and the total work done along curve C is:

$$\sum \mathbf{F(R)} \bullet \Delta\mathbf{R}$$

Taking the limit gives the work done by **F(R)** along curve C:

$$\lim\nolimits_{|\Delta\mathbf{R}| \to 0} \sum \mathbf{F(R)} \bullet \Delta\mathbf{R} = \int_C \mathbf{F} \bullet d\mathbf{R} = \int_C (F_1 dx + F_2 dy + F_3 dz)$$

where:

F₁dx is (force in x-direction)(movement in x-direction)

F₂dy is (force in y-direction)(movement in y-direction)

F₃dz is (force in z-direction)(movement in z-direction)

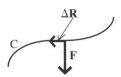

• Consider the work around a closed curve described by C_1 and C_2, where C_1 is a half circle from 0 to π of radius 1 and C_2 is a straight line from -1 to 1. The motion occurs counter-clockwise. What is $\int_C \mathbf{F} \bullet d\mathbf{R}$ for C_1 and C_2?

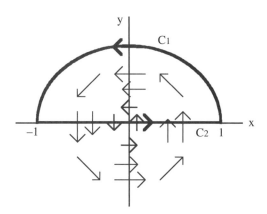

F is given by: $\mathbf{F} = -y\mathbf{i} + x\mathbf{j}$, where $x = \cos t$, $y = \sin t$ and
$(x(t), y(t)) = (\cos t, y \sin t)$ are parameters.

$\mathbf{R} = x\mathbf{i} + y\mathbf{j} = (\cos t)\mathbf{i} + (\sin t)\mathbf{j}$

Therefore, $d\mathbf{R} = (-\sin t)\mathbf{i} + (\cos t)\mathbf{j}$.

For C_1, $\int_C \mathbf{F} \cdot d\mathbf{R} = {_0}\!\int^\pi (-\sin t\,\mathbf{i} + \cos t\,\mathbf{j}) \cdot (-\sin t\,\mathbf{i} + \cos t\,\mathbf{j})dt$

$= {_0}\!\int^\pi (\sin^2 t + \cos^2 t)dt = \pi$

For C_2, $\mathbf{F} = -y\mathbf{i} + x\mathbf{j} = \mathbf{F} = 0 + x\mathbf{j}$. Therefore:

$\int_C \mathbf{F} \cdot d\mathbf{R} = {_0}\!\int^1 (\cos t\,\mathbf{j}) \cdot (-\sin t\,\mathbf{i} + \cos t\,\mathbf{j})dt$

Because **F** does not have an **i** component on the X-axis where $y = 0$ and **F** is perpendicular to C_2 along the length of C_2, then $\mathbf{F} \cdot d\mathbf{R} = 0$

Therefore, C_1 and C_2 combine to:

$\int_C \mathbf{F} \cdot d\mathbf{R} = \int_{C1} \mathbf{F} \cdot d\mathbf{R} + \int_{C2} \mathbf{F} \cdot d\mathbf{R} = \pi + 0 = \pi$

• In general, a *line integral* over curve C from point a to point b depends on points a and b as well as the *path of the curve*. There are, however, vector fields such as gradient fields where the line integral does not depend on the path of the curve but only on the beginning and ending points. A *line integral is independent of path* when the value of the integral is the difference of the values of f at the beginning and ending points of C, where C is the path from point a to point b. The position vector is:

$\mathbf{R}(t) = x(t)\mathbf{i} + y(t)\mathbf{j} + z(t)\mathbf{k}$, $a \leq t \leq b$, and the integral is:

${_a}\!\int^b (F_1 dx + F_2 dy + F_3 dz) =$

${_a}\!\int^b ((\partial f/\partial x)dx + (\partial f/\partial y)dy + (\partial f/\partial z))dz = {_a}\!\int^b (\partial f/\partial t)dt = f(B) - f(A)$

• In *conservative vector fields*, which are *gradient fields*, all paths of integration result in the same value of work done. (Remember that every gradient field is conservative.) If **F** is a conservative vector field, energy is conserved and no work is done in the displacement of an object from point P back to point P. In a conservative vector field, if a body moves from a starting point back to the starting point, when it returns to its starting point it will have the same kinetic energy it had originally.

• Line integrals around *closed curves* or closed paths are not always zero. However, the *line integral* for a closed path is independent of path if its value is zero. In the figure:

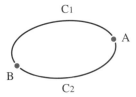

$\int_C \mathbf{F} \bullet d\mathbf{R}$ can be *independent of path* if, when integrating from A to B along C_1 then from B to A along C_2 (in the -1 direction), the sum of these two integrals must be zero.

• In general, in a *conservative gradient vector* field where

   $\mathbf{F} = \text{grad } f$:

(a.) $\int_C (\text{grad } f) \bullet d\mathbf{R} = f(P_2) - f(P_1)$ for curve C between points P_1 and P_2, where the work depends on the beginning and ending points rather than the path.

(b.) $\int_C \mathbf{F} \bullet d\mathbf{R}$ has the same value along any path from point P_1 to point P_2.

(c.) The work $\int_C \mathbf{F} \bullet d\mathbf{R}$ around every *closed path* is zero.

(d.) The components satisfy $(\partial F_2/\partial x) = (\partial F_1/\partial y)$.

7.4 Green's Theorem: Tangent and Normal (Flux) Forms

• This section includes Green's Theorem in its tangent form, applying it in a vector field, the development of an expression for area, Green's Theorem in its normal (or flux) form, a comparison of the two forms, and Green's Theorem in vector fields that are conservative and source-free.

• *Green's Theorem* connects *line integrals* with *surface integrals*. In its tangent form, Green's Theorem relates *work* to *curl* (see Section 7.7 for curl) and in its normal form Green's Theorem relates *flux* to *divergence*. (See Section 7.6 for divergence.)

- For *circulation or movement around a curve* C enclosing a region, *Greens Theorem* (in tangent form) connects a *double integral* over region R to a *line integral* along its boundary C. If R is a closed region in an XY plane bound by curve C (which consists of many smooth curves and does not cross itself) then the integral around C equals the integral over R. Therefore, Green's Theorem in its *tangential form along* C enclosing region R is:

$$\oint_C [F_1 dx + F_2 dy] = \iint_R [(\partial F_2/\partial x) - (\partial F_1/\partial y)]dxdy,$$

or equivalently:

$$\oint_C [\, \mathbf{F} \bullet d\mathbf{R}] = \iint_R [(\partial F_2/\partial x) - (\partial F_1/\partial y)]dxdy$$

(work = curl)

where $F_1(x,y)$ and $F_2(x,y)$ are functions that are continuous and have continuous partial derivatives $(\partial F_2/\partial x)$ and $(\partial F_1/\partial y)$ everywhere in the domain containing region R.

- If $\mathbf{F} = F_1 \mathbf{i} + F_2 \mathbf{j}$ is a *gradient field*, it has a potential function f and the property $(\partial F_2/\partial x) = (\partial F_1/\partial y)$. Therefore:

$$\oint_C [\, \mathbf{F} \bullet d\mathbf{R}] = \iint_R [(\partial F_2/\partial x) - (\partial F_1/\partial y)]dxdy = 0$$

Therefore, if $\mathbf{F}$ is a conservative field, $(\partial F_2/\partial x) = (\partial F_1/\partial y)$ and work is zero.

- Consider the domain containing a region where vector field $\mathbf{F} = F_1 \mathbf{i} + F_2 \mathbf{j}$ is located and is assumed to have no holes and every point is enclosed by curve C.

If C is a circle of radius 1 centered at the origin and:

$$\mathbf{F} = \frac{-y\mathbf{i} + x\mathbf{j}}{x^2 + y^2}, \text{ where } F_1 = \frac{-y}{x^2 + y^2}, F_2 = \frac{x}{x^2 + y^2}$$

then along C, $\mathbf{F}$ is tangent to the circle of radius 1, $|\mathbf{F}| = 1$, and $d\mathbf{R}$ is the length of the curve, which is 2π, (the circumference with $r = 1$). Therefore:

$$\oint_C [\, \mathbf{F} \bullet d\mathbf{R}] = 1 \bullet 2\pi = 2\pi$$

If **F** is a gradient field, demonstrate that $(\partial F_2/\partial x) = (\partial F_1/\partial y)$. Using the product rule: $(f/g)' = (f'g - fg')/g^2$:

$$\frac{\partial F_1}{\partial y} = \frac{\partial}{\partial y}\frac{-y}{(x^2+y^2)} = \frac{-1(x^2+y^2)+y(2y)}{(x^2+y^2)^2} = \frac{-x^2-y^2+2y^2}{(x^2+y^2)^2}$$

$$= \frac{y^2-x^2}{(x^2+y^2)^2}$$

$$\frac{\partial F_2}{\partial x} = \frac{\partial}{\partial x}\frac{x}{(x^2+y^2)} = \frac{1(x^2+y^2)-x(2x)}{(x^2+y^2)^2} = \frac{x^2+y^2-2x^2}{(x^2+y^2)^2}$$

$$= \frac{y^2-x^2}{(x^2+y^2)^2}$$

Therefore, $(\partial F_2/\partial x) = (\partial F_1/\partial y)$.

Also note that at $x = 0$, $y = 0$, $(\partial F_2/\partial x)$ and $(\partial F_1/\partial y)$ do not exist and, therefore, Green's Theorem does not hold true for any region containing the origin.

• *Green's Theorem* can be used to develop an expression for the calculation of *area* of a region.

$$\oint_C [F_1 dx + F_2 dy] = \iint_R [(\partial F_2/\partial x) - (\partial F_1/\partial y)]dxdy$$

When $F_1 = 0$, $F_2 = x$, Green's Theorem reduces to:

$$\oint_C x\, dy = \iint_R dxdy$$

When $F_1 = -y$, $F_2 = 0$, Green's Theorem reduces to:

$$- \oint_C y\, dx = \iint_R dxdy$$

where $\iint_R dxdy$ = area A. Adding the two expressions above gives:

$$2\iint_R dxdy = \oint_C (x\, dy - y\, dx)$$

or

$$\text{Area} = A = \iint_R dxdy = (1/2)\oint_C (x\, dy - y\, dx)$$

- The equation of an ellipse is given by $\dfrac{x^2}{a^2} + \dfrac{y^2}{b^2} = 1$ or equivalently, $x = a \cos t$, $y = b \sin t$. Find area using

$$(1/2) \oint_C (x\,dy - y\,dx).$$

If the points on the ellipse are (x,y) as t goes from 0 to 2π, and $dx = -a \sin t$ and $dy = b \cos t$, Then area is:

$$A = (1/2)\,_0\!\int^{2\pi} (x\,dy - y\,dx)$$

$$= (1/2)\,_0\!\int^{2\pi} (ab \cos^2 t - (-ab \sin^2 t))dt = \pi ab$$

- *Green's Theorem across a curve (flux)* gives *Green's Theorem in its normal form.* Consider a *flow field* $\mathbf{F} = F_1(x,y)\mathbf{i} + F_2(x,y)\mathbf{j}$, with steady flow across boundary C, where (flow out – flow in) is balanced by a replacement of fluid in the side of region R. The normal form of Green's Theorem for flux *across* C enclosing region R is:

$$\oint_C [F_1 dy - F_2 dx] = \iint_R [(\partial F_1/\partial x) + (\partial F_2/\partial y)]dxdy$$

(flux = divergence)

- The following is a *comparison of the flux form of Green's Theorem with the tangent form*:

$\oint_C$ in tangent form is: $\oint_C [F_1 dx + F_2 dy]$, which is *work*.

$\oint_C$ in normal form is: $\oint_C [F_1 dy - F_2 dx]$, which is *flux*.

Also:

$\iint_R$ in tangent form is: $\iint_R [(\partial F_2/\partial x) - (\partial F_1/\partial y)]dxdy$

which is *curl*.

$\iint_R$ in normal form is: $\iint_R [(\partial F_1/\partial x) + (\partial F_2/\partial y)]dxdy$

which is *divergence*.

Note that the divergence of a flow field is $(\partial F_1/\partial x) + (\partial F_2/\partial y)$. See Sections 7.5 for flux, 7.6 for divergence, and 7.7 for curl.

• The total *flow across a defined region*, such as a rectangle in a coordinate system, can be depicted as:

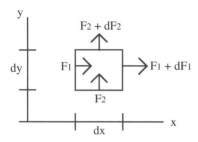

Flow from left to right through the rectangle is given by (change in F_1)(dy). Flow from bottom to top through the rectangle is given by (change in F_2)(dx). *Total flow out of rectangle* is:
$$dF_1 dy + dF_2 dx = [(\partial F_1/\partial x) + (\partial F_2/\partial y)]dxdy$$

Therefore, the *divergence* multiplied by area dydx is the total flow out.

• In general, a *flow field in region R* balances flow through curve C (flow out – flow in) with a replacement in region *R* (source – sink). In a *flow field without a source*, the flux is zero through C and the divergence is also zero.

$(\partial F_1/\partial x) + (\partial F_2/\partial y) = 0$. In the *source-free field*:

$\mathbf{F} = F_1(x,y)\mathbf{i} + F_2(x,y)\mathbf{j}$, the flux is:

$\oint_C \mathbf{F} \cdot \mathbf{n}ds$ through every closed curve is zero, and also

$\oint_C \mathbf{F} \cdot \mathbf{n}ds$ between any two points is the same.

Also, in a source-free field, a *stream function* g exists and is described in terms of $F_1 = (\partial g/\partial y)$ and $F_2 = -(\partial g/\partial x)$.

• In summary, if a *vector flow field* **F** *is conservative and source-free*, then *curl* $\mathbf{F} = (\partial F_2/\partial x) - (\partial F_1/\partial y)$ is zero and *divergence* $\mathbf{F} = (\partial F_1/\partial x) + (\partial F_2/\partial y)$ is zero.

Because the *field is conservative*, there exists a potential f where $F_1 = (\partial f/\partial x)$ and $F_2 = (\partial f/\partial y)$.

Because the *field is source-free*, there is a stream function g where $F_1 = (\partial g/\partial y)$ and $F_2 = -(\partial g/\partial x)$.

Therefore, when field **F** is both conservative and source-free,
$(\partial F_1/\partial y) = (\partial F_2/\partial x)$, $(\partial F_1/\partial x) = -(\partial F_2/\partial y)$ and $F_1 = (\partial f/\partial x) = (\partial g/\partial y)$
and $F_2 = (\partial f/\partial y) = -(\partial g/\partial x)$.
(These are called the *Cauchy-Riemann equations*.)
Also, there exists a *potential function* f and a *stream function* g and the
Laplace's equations are satisfied:

$$(\partial^2 f/\partial x^2) + (\partial^2 f/\partial y^2) = (\partial F_1/\partial x) + (\partial F_2/\partial y) = 0$$
$$(\partial^2 g/\partial x^2) + (\partial^2 g/\partial y^2) = -(\partial F_2/\partial x) + (\partial F_1/\partial y) = 0$$

7.5 Surface Integrals and Flux

• This section includes flux through a surface and the surface integral,
examples of flux in various vector fields, and examples of calculating flux.

• *Flux* represents the rate of flow or movement through a surface. For
example, in a velocity vector field, flux represents the rate of fluid flow
through a surface, or the volume of fluid that crosses a surface per unit
time. To evaluate *flux* across a surface, such as mass or fluid crossing a
surface in a given time period, a *flux integral* can be used.

• A flux integral over surface S is a *surface integral* of a vector function **F**
and can be written as:

$$\iint_S \mathbf{F} \cdot \mathbf{n}\, dS$$

where $\iint dS$ is area, $\mathbf{F} \cdot \mathbf{n}$ is the normal component of **F**. The expression
$\mathbf{F} \cdot \mathbf{n}\, dS$ is also written $\mathbf{F} \cdot d\mathbf{S}$, where $\mathbf{n}dS = d\mathbf{S}$ is a vector with direction
n and magnitude dS. Similarly, the expression $\mathbf{F} \cdot \mathbf{n}\, dA$ is also written
$\mathbf{F} \cdot d\mathbf{A}$, where $\mathbf{n}dA = d\mathbf{A}$.

• The flux integral over a surface using parameters u and v can be written:

$$\iint_S \mathbf{F} \cdot \mathbf{n}\, dS = \iint_R \mathbf{F}(\mathbf{R}(u,v)) \cdot \mathbf{n}(u,v)\, dudv$$

where *R* is the region in the uv-plane that corresponds to surface S (where
the surface is projected), $\mathbf{N}\, dudv = \mathbf{n}|\mathbf{N}|$ and $|\mathbf{N}| = |\mathbf{Ru} \times \mathbf{Rv}|$ is the
area of the parallelogram with sides **Ru** and **Rv**. The direction of flow is
$\mathbf{n} = \mathbf{N}/|\mathbf{N}|$.

• The flux through a surface can be positive or negative depending on the direction of flow or the choice of direction or orientation of the surface. Because flux through a surface is dependent on the direction the surface faces as well as the area of surface, it is advantageous to represent the area as a vector quantity.

• Consider the simple case of *constant flow* through a pipe where the flux through a defined circular region is:

(flow rate)(area of region)

If the flow is *variable*, the surface can be sectioned into small areas where flow is approximately constant in each section and represented as the limit of the sum:

$$\lim_{|\Delta A| \to 0} \sum V \bullet \Delta A = \iint V \bullet dA$$

where **V** is the velocity vector.

Flux through a surface can be applied to any vector field **F** not only a velocity field.

• In general, *flux through a curved surface* can be thought of as the sum of the fluxes through many small almost flat sections that the surface is divided into. The small, almost flat sections are called parameter rectangles Δx and Δy, which align with X- and Y-axes.

$$\text{Flux} = \lim_{|\Delta A| \to 0} \sum F \bullet \Delta A = \iint F \bullet dA$$

Where the limit exists, **F** is continuous in the region containing the surface and the subsections that the surface is divided into are smooth.

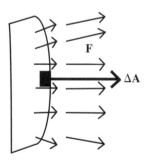

• Following are examples of *flux in vector fields*:

(a.) Radial field:

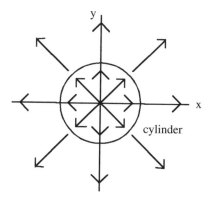

This is a radial field in the XY plane containing a cylinder with the Z-axis pointing out of the page in the direction of the cylinder. The radial field points outward everywhere along the Z-axis. The area across the ends of the cylinder has no flux because the flow is parallel. The flux is $\iint \mathbf{F} \cdot d\mathbf{S}$ and the flow is normal to the surface of the cylinder.

(b.) Rotation field:

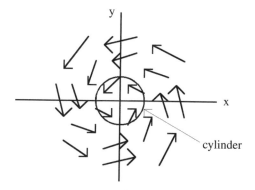

This example is similar to (a.) except the flow is not parallel to the normal vector of the cylinder and is rotating or spinning in a slightly inward direction. The flux integral is negative.

(c.) Horizontal field:

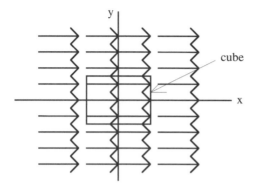

This vector field is parallel to the X-axis through a cube. The faces of the cube that are parallel to the flow and to the X-axis have a flux of zero. The flux through the two faces perpendicular to flow are equal in magnitude and opposite in sign therefore the net flux is zero.

(d.) Field through a sphere:

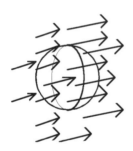

This is a closed spherical surface in flow field **F** oriented with the positive direction of flow from inside to outside. Area vectors d**S** of the sphere all point outward. The flux through the surface is $\iint \mathbf{F} \cdot d\mathbf{S}$, which is the flux out of the region enclosed by the surface.

(e.) Radial field out of a sphere:

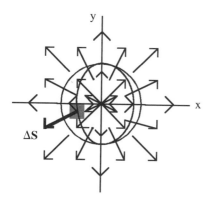

This radial vector field $\mathbf{V}$ points in the same direction as the surface normal vector $\Delta\mathbf{S}$. The sphere has a radius r where $r = |\mathbf{V}|$ on the surface. Therefore:

$$\mathbf{V} \cdot \Delta\mathbf{S} = |\mathbf{V}||\Delta\mathbf{S}| = r|\Delta\mathbf{S}|$$

Summing over all sections on the surface and taking the limit:

$$\lim_{|\Delta\mathbf{S}| \to 0} \sum \mathbf{V} \cdot \Delta\mathbf{S} = \lim_{|\Delta\mathbf{S}| \to 0} \sum r|\Delta\mathbf{S}| = r[\lim_{|\Delta\mathbf{S}| \to 0} \sum |\Delta\mathbf{S}|]$$

which is the surface area of the sphere multiplied by the radius. The flux out of this radial field is:

$$\text{Flux} = \iint \mathbf{V} \cdot d\mathbf{S} = r[\lim_{|\Delta\mathbf{S}| \to 0} \sum |\Delta\mathbf{S}|] = r(4\pi r^2) = 4\pi r^3$$

• To develop an expression for *flux*, consider a surface that is sectioned into small *parameter rectangles*. It is useful to remember the *area of a parallelogram* (discussed in Section 5.5). Remember that two vectors $\mathbf{A}$ and $\mathbf{B}$ form a parallelogram and the length of the vector $\mathbf{C}$ resulting from their cross product is the area of the parallelogram. Vector $\mathbf{C}$ is the vector normal $\mathbf{N}$ to the surface.

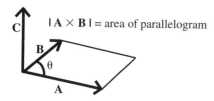

$|\mathbf{A} \times \mathbf{B}| = $ area of parallelogram

In a surface $z = f(x,y)$, the area of each *parameter rectangle* is the cross product of its sides. Therefore, the area vector for a parameter rectangle on the surface $z = f(x,y)$ can be represented by the cross product of its sides using position vectors. If the parameter rectangle is located at point P, it can be represented with *position vector*:

$$\mathbf{R} = x\mathbf{i} + y\mathbf{j} + z\mathbf{k} = x\mathbf{i} + y\mathbf{j} + f(x,y)\mathbf{k}$$

At point P, two parameter curves x_0 and y_0 cross (see figure below). Vectors tangent to curves x_0 and y_0 at point P are:

$$(\partial \mathbf{R}/\partial x) = \mathbf{i} + (\partial f(x_0,y_0)/\partial x)\mathbf{k} = \mathbf{i} + (\partial z/\partial x)\mathbf{k}$$

which is the change at $y = y_0$.

$$(\partial \mathbf{R}/\partial y) = \mathbf{j} + (\partial f(x_0,y_0)/\partial y)\mathbf{k} = \mathbf{j} + (\partial z/\partial y)\mathbf{k}$$

which is the change at $x = x_0$.

The position vector $\mathbf{R}$ along curve y_0 changes in the x-direction:

$$\Delta \mathbf{R} \approx \Delta x\mathbf{i} + (\partial f(x_0,y_0)/\partial x)\Delta x\mathbf{k} = (\partial \mathbf{R}/\partial x)\Delta x$$

The position vector $\mathbf{R}$ along curve x_0 changes in the y-direction:

$$\Delta \mathbf{R} \approx \Delta y\mathbf{j} + (\partial f(x_0,y_0)/\partial y)\Delta y\mathbf{k} = (\partial \mathbf{R}/\partial y)\Delta y$$

Therefore, the area of the parameter rectangle $\Delta \mathbf{S}$ on a surface at point P can be represented as the cross product of the sides:

$$(\partial \mathbf{R}/\partial x)\Delta x \times (\partial \mathbf{R}/\partial y)\Delta y = \begin{bmatrix} \mathbf{i} & \mathbf{j} & \mathbf{k} \\ 1 & 0 & \partial f/\partial x \\ 0 & 1 & \partial f/\partial y \end{bmatrix} \Delta x \Delta y$$

$$= \Delta \mathbf{S} = ((-\partial f(x_0,y_0)/\partial x)\mathbf{i} - (\partial f(x_0,y_0)/\partial y)\mathbf{j} + \mathbf{k})\Delta x \Delta y$$

The *flux* through a surface $z = f(x,y)$ in the positive z-direction can be written as this general expression:

$$\iint_S \mathbf{F} \bullet d\mathbf{S} = \iint_R \mathbf{F}(x,y,f(x,y)) \bullet ((-\partial f/\partial x)\mathbf{i} - (\partial f/\partial y)\mathbf{j} + \mathbf{k})dxdy$$

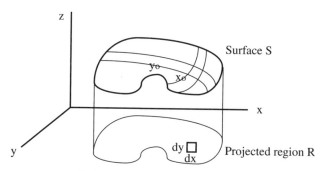

Projection of parameter rectangle

• Example: What is the flux through a cone where $z = \sqrt{x^2 + y^2}$ and $\mathbf{F} = x\mathbf{i} + y\mathbf{j} + z\mathbf{k}$ is a radial field?

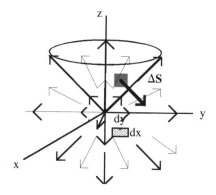

The height of the cone is z and:

$$(\partial z/\partial x) = x / \sqrt{x^2 + y^2}$$

$$(\partial z/\partial y) = y / \sqrt{x^2 + y^2}$$

Using the expression for $\Delta\mathbf{S}$ or:

$$d\mathbf{S} = (-\partial f/\partial x)\mathbf{i} - (\partial f/\partial y)\mathbf{j} + \mathbf{k}$$

$$= (-x / \sqrt{x^2 + y^2}\,)\mathbf{i} - (y / \sqrt{x^2 + y^2}\,)\mathbf{j} + \mathbf{k} = \mathbf{n}dS$$

The flux is:

$$\iint_S \mathbf{F} \cdot d\mathbf{S} =$$

$$\iint_R (x\mathbf{i} + y\mathbf{j} + \sqrt{x^2+y^2}\ \mathbf{k}) \cdot (\frac{-x}{\sqrt{x^2+y^2}}\,\mathbf{i} - \frac{y}{\sqrt{x^2+y^2}}\,\mathbf{j} + \mathbf{k})\ dS$$

$$= \iint (\frac{-x^2}{\sqrt{x^2+y^2}} - \frac{y^2}{\sqrt{x^2+y^2}} + \sqrt{x^2+y^2}\ )\ dS$$

$$= \iint (\frac{-x^2-y^2}{\sqrt{x^2+y^2}} + \frac{x^2+y^2}{\sqrt{x^2+y^2}})\ dS = \iint (0)\ dS = 0$$

The flux is zero because $\mathbf{F}$ is parallel to the sides of the cone, and the normal vector to the surface is perpendicular to $\mathbf{F}$. Therefore, there is no flow through the sides of the cone because $\mathbf{F} \cdot \mathbf{n} = 0$.

- *Flux* through a surface using *parameters* u and v in field $\mathbf{F} = F_1\mathbf{i} + F_2\mathbf{j} + F_3\mathbf{k}$ is given by:

$$\iint \mathbf{F} \cdot \mathbf{n}\,dS = \iint \mathbf{F} \cdot \mathbf{N}\ dudv = \iint \mathbf{F} \cdot (\mathbf{A} \times \mathbf{B})\ dudv$$

where a small section of the surface has area:

$$dS = |\mathbf{A} \times \mathbf{B}|\,dudv = |\mathbf{N}|\,dudv.$$

$\mathbf{A}$ and $\mathbf{B}$ are the vectors along the side of dS where:

$$\mathbf{A} = (\partial x/\partial u)\mathbf{i} + (\partial y/\partial u)\mathbf{j} + (\partial z/\partial u)\mathbf{k}$$
$$\mathbf{B} = (\partial x/\partial v)\mathbf{i} + (\partial y/\partial v)\mathbf{j} + (\partial z/\partial v)\mathbf{k}$$
$$\mathbf{A} \times \mathbf{B} = \mathbf{N}, \text{ and the unit normal vector is } \mathbf{n} = \mathbf{N}/|\mathbf{N}|.$$

- **Example:** Find flux through a surface in velocity field $\mathbf{F} = y\mathbf{i} + 2\mathbf{j} + xz\mathbf{k}$ where the surface is given by $y = x^2$ from $0 \le x \le 2$, $0 \le z \le 1$.

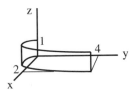

Parameters x = u, y = x^2 = u^2, z = v, can be used to represent S:

$\mathbf{R}$ = u$\mathbf{i}$ + u$^2\mathbf{j}$ + v$\mathbf{k}$

$(\partial\mathbf{R}/\partial u)$ = $\mathbf{i}$ + 2u$\mathbf{j}$

$(\partial\mathbf{R}/\partial v)$ = $\mathbf{k}$

$\mathbf{N}$ = $(\partial\mathbf{R}/\partial u)$ × $(\partial\mathbf{R}/\partial u)$ = (2u − 0)$\mathbf{i}$ + (0 − 1)$\mathbf{j}$ + (0 − 0)$\mathbf{k}$ = 2u$\mathbf{i}$ − $\mathbf{j}$

Substituting for $\mathbf{F}$:

$\mathbf{F}$(u,v) = u$^2\mathbf{i}$ + 2$\mathbf{j}$ + uv$\mathbf{k}$

Therefore:

$\mathbf{F} \cdot \mathbf{N}$ = (u$^2\mathbf{i}$ + 2$\mathbf{j}$ + uv$\mathbf{k}$) • (2u$\mathbf{i}$ − $\mathbf{j}$) = 2u^3 − 2

Integrate using parameters, 0 ≤ u ≤ 2, 0 ≤ v ≤ 1:

$\iint_S \mathbf{F} \cdot \mathbf{n} dS = {_0}\int^1 {_0}\int^2 (2u^3 - 2)\ du\ dv = {_0}\int^1 (4)\ dv = 4\ \text{units}^3/\text{time}$

7.6 Divergence

• This section includes the definition and notation for divergence, the Divergence Theorem, and examples of divergence in vector fields.

• *Divergence* represents the strength of outflow from a *point* in a vector field. In a velocity field, the divergence gives the outflow per unit volume at a point. In fluid flow, divergence is the rate at which mass leaves an enclosed region in $\mathbf{F}$, or *flux per unit volume*. The *Divergence Theorem* relates *divergence* to *flux*.

• Remember that flux represents the net outflow of, for example, fluid through a surface surrounding a region in a vector field (e.g., a velocity field of incompressible fluid). Whereas the *divergence of a vector field* represents the outflow per unit volume at a point.

• The *divergence of vector field* $\mathbf{F}$ is given by:

div $\mathbf{F}$ = $(\partial F_1/\partial x)$ + $(\partial F_2/\partial y)$ + $(\partial F_3/\partial z)$

where $\mathbf{F}(x,y,z)$ is a differentiable vector function, F_1, F_2, and F_3 are the components of $\mathbf{F}$, and x, y, and z are Cartesian coordinates. Note that the value of (div $\mathbf{F}$) does not depend on the coordinate system used but on the points in space.

• Divergence is commonly written in the forms:

$\mathrm{div}\ \mathbf{F} = \nabla \bullet \mathbf{F}$

$= [(\partial/\partial x)\mathbf{i} + (\partial/\partial y)\mathbf{j} + (\partial/\partial z)\mathbf{k}] \bullet [F_1\mathbf{i} + F_2\mathbf{j} + F_3\mathbf{k}]$

$= (\partial F_1/\partial x) + (\partial F_2/\partial y) + (\partial F_3/\partial z)$

where ∇ is the *del* operator and is given by:

$(\partial/\partial x) + (\partial/\partial y) + (\partial/\partial z)$

• The divergence of a vector field, div $\mathbf{F}$ (or $\nabla \bullet \mathbf{F}$), is a scalar valued function, whereas (grad f) or ∇f results in a vector.

• **Example:** If $\mathbf{F} = x^2z\mathbf{i} + xy\mathbf{j} + yz^2\mathbf{k}$, then the divergence is:

$\mathrm{div}\ \mathbf{F} = 2xz + x + 2yz$

• To visualize divergence, consider a small rectangular region at point P in a velocity vector field $\mathbf{F}(x,y,z)$. In this velocity field, the motion of fluid in the region has no sources or sinks; therefore, no fluid is generated or consumed. The dimensions of this region are Δx, Δy, Δz, with the edges parallel to the coordinate axes and the volume of the region is $\Delta x \Delta y \Delta z = \Delta V$.

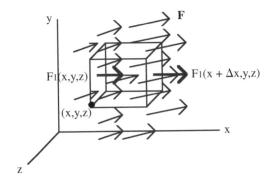

The rectangular region is small enough so that its sides are approximately flat and $\mathbf{F}$ is approximately constant on each face.

The flux in the direction along the X-axis is perpendicular to the left face and is approximately equal to the x-component of $\mathbf{F}$ multiplied by the area of that face: $F_1(x,y,z)\Delta y\Delta z$.

The flux along the X-axis leaving the region is perpendicular to the right face and is approximately equal to the $(x + \Delta x)$ component of $\mathbf{F}$ multiplied by the area of the face: $F_1(x+\Delta x,y,z)\Delta y\Delta z$.

Therefore, the net flux out of this region along the X-axis is:

$$F_1(x+\Delta x,y,z)\Delta y\Delta z - F_1(x,y,z)\Delta y\Delta z$$
$$= [F_1(x+\Delta x,y,z) - F_1(x,y,z)]\Delta y\Delta z\Delta x/\Delta x = (\partial F_1/\partial x)\Delta x\Delta y\Delta z$$

Similarly, the net flux in the y-direction perpendicular to the top and bottom faces is:

$$F_2(y+\Delta y,x,z)\Delta x\Delta z - F_2(x,y,z)\Delta x\Delta z = (\partial F_2/\partial y)\Delta x\Delta y\Delta z$$

Similarly, the net flux in the z-direction perpendicular to the front and back faces is:

$$F_3(z+\Delta z,x,y)\Delta x\Delta y - F_3(x,y,z)\Delta x\Delta y = (\partial F_3/\partial z)\Delta x\Delta y\Delta z$$

Therefore, the net flux out of the region is:

$$(\partial F_1/\partial x)\Delta x\Delta y\Delta z + (\partial F_2/\partial y)\Delta x\Delta y\Delta z + (\partial F_3/\partial z)\Delta x\Delta y\Delta z$$

The net outflow per volume of the region where volume $\Delta V = \Delta x\Delta y\Delta z$ is:

$$\text{div } \mathbf{F} = (\partial F_1/\partial x) + (\partial F_2/\partial y) + (\partial F_3/\partial z)$$

• If a vector field represents flow away from a point, the divergence is ≥ 0. Conversely, if a vector field represents flow toward a point, the divergence is ≤ 0. Therefore, divergence of $\mathbf{F}$ essentially measures the *source*, because

(flow out of a region) – (flow into a region) = source.

If a vector field $\mathbf{F}$ has zero divergence at every point, it is called *divergence-free*.

• If *divergence is not a constant value*, the flux out of the total volume is represented using the sum of sections that make up the total volume. The divergence in each section is nearly constant and the flux out of each section is approximately: div $\mathbf{F}(x,y,z)\Delta V$.

If all the sections within the total volume are summed, the flux out of the total volume is:

Σ(flux out of each section) $\approx \Sigma$ [div $\mathbf{F}(x,y,z)\Delta V$].

As the size of each section approaches zero, the sum becomes:

flux out of total volume $= \iiint_V$ div $\mathbf{F}$ dV

Therefore, the *flux* of a vector field $\mathbf{F}$ through a *closed surface* can be represented by using flux integral

$\iint_S \mathbf{F} \bullet d\mathbf{A}$, or using the integral of divergence

$\iiint_V$ div $\mathbf{F}$ dV

• *Gauss's Divergence Theorem* relates *surface integrals* to *triple integrals*. For region R in space closed and bounded by a piece-wise smooth surface S, if $\mathbf{F}(x,y,z)$ is a vector function that is continuous and has continuous first partial derivatives in the domain containing R, then:

$\iiint_R$ div $\mathbf{F}$ dV $= \iint_S \mathbf{F} \bullet \mathbf{n}$ dS

where R represents the volume enclosed by the surface S and $\mathbf{n}$ is the outer normal vector of S.

• Divergence can be written without reference to coordinates by dividing the Divergence Theorem in the form:

$\iiint_R$ div $\mathbf{F}$ dV $= \iint_S \mathbf{F} \bullet \mathbf{n}$ dA

by the volume of region R that is enclosed by surface S:

$(1/V(R)) \iiint_R$ div $\mathbf{F}$ dV $= (1/V(R)) \iint_S \mathbf{F} \bullet \mathbf{n}$ dA

where $\mathbf{n}$ is the outer unit normal vector of S.

In general, the divergence of vector field $\mathbf{F}$ at a point P can be defined by:

div $\mathbf{F}(P) = \lim_{V \to 0}[(1/\text{volume enclosed by S})\iint_S \mathbf{F} \bullet d\mathbf{A}]$

where S is the surface that encloses point P such that the volume V inside S approaches zero.

• Consider steady flow of incompressible fluid in a velocity field **V** where the density is constant and equal to 1. Region R is bounded by surface S where **n** is the unit normal vector pointing out of the surface. The total mass of fluid moving outward across S from region R per unit time is the total flow out of R:

$$\iint_S \mathbf{V} \cdot \mathbf{n} \, dA$$

where dA is the area of each small section of the surface.

The *average flow out* of R is:

$$(1/V) \iint_S \mathbf{V} \cdot \mathbf{n} \, dA$$

where V is the volume of R.

For steady flow of an incompressible fluid, the flow out of the region must be replaced continuously if the above integral is not zero. In this case, there must exist sources or sinks within R where fluid is produced or consumed.

If R gets smaller until it is on some point P in R, then the source intensity at point P is:

$$\lim_{dR \to 0} (1/V(R)) \iint_{S(R)} \mathbf{V} \cdot \mathbf{n} \, dA$$

Therefore, the divergence of the velocity vector **V** for a steady, incompressible flow is the source intensity of the flow at that corresponding point.

If there are no sources in R, then:

$$\text{div } \mathbf{V} = 0 \text{ and } \iint_{S(R)} \mathbf{V} \cdot \mathbf{n} \, dA = 0$$

• **Example:** In *radial vector field* **V** (discussed in vector field example (e.) in the Section 7.5), the flux across a sphere of radius r centered at the origin of the radial field is $4\pi r^3$. The average outflow or flux per unit volume at a point in the sphere is: flux/volume of sphere = $4\pi r^3/(4/3)\pi r^3 = 3$ cubic units of flow per unit time per unit space.

The flux or outflow per unit volume at the origin is the limit as radius r approaches zero of the (flux/volume). This limit also results in the value 3, which is the divergence of **V** at the origin:

$$\text{div } \mathbf{V} = \text{div}(x\mathbf{i} + y\mathbf{j} + z\mathbf{k}) = (\partial x/\partial x) + (\partial y/\partial y) + (\partial z/\partial z)$$
$$= 1 + 1 + 1 = 3$$

Therefore, if **V** is a velocity field consisting of an incompressible fluid, then fluid is created at 3 units fluid/unit volume at all points and the total fluid production in the sphere is:

$3(4/3)\pi r^3 = 4\pi r^3$, which is the flux.

7.7 Curl

• This section includes the definition of curl and curl in various vector fields.

• *Curl of a vector field* measures the strength of *rotation* or spin around a point. Remember that *divergence* measures the flow away from or toward a point. The curl of a vector field at a point is the vector pointing in the direction of maximum *circulation* strength, and the magnitude of curl is the strength of the circulation. For example, for a rigid body, curl measures rotation or spin, the direction of curl points in the axis of rotation and the magnitude of curl is two-times the speed of rotation. The curl of a vector field is itself a vector field.

• Remember that the *gradient* gives the direction of greatest increase, such that the maximum increase of f is $|\text{grad } f|$ in the direction of (grad f). Similarly, *curl* gives the direction of maximum rotation, such that maximum rotation rate of **F** is $(1/2)|\text{curl } \mathbf{F}|$ in the direction of curl **F**.

• The *curl of a vector field* $\mathbf{F}(x,y,z) = F_1\mathbf{i} + F_2\mathbf{j} + F_3\mathbf{k}$ is the vector field given by:

$$\text{curl } \mathbf{F} = \nabla \times \mathbf{F} = \begin{vmatrix} \mathbf{i} & \mathbf{j} & \mathbf{k} \\ \partial/\partial x & \partial/\partial y & \partial/\partial z \\ F_1 & F_2 & F_3 \end{vmatrix}$$

$$= \left(\frac{\partial F_3}{\partial y} - \frac{\partial F_2}{\partial z}\right)\mathbf{i} + \left(\frac{\partial F_1}{\partial z} - \frac{\partial F_3}{\partial x}\right)\mathbf{j} + \left(\frac{\partial F_2}{\partial x} - \frac{\partial F_1}{\partial y}\right)\mathbf{k}$$

This is referred to as the curl of vector function **F** or equivalently the curl of the vector field defined by **F**.

- In a two-dimensional planar vector field, where:

$$\mathbf{F}(x,y) = F_1(x,y)\mathbf{i} + F_2(x,y)\mathbf{j}, \text{ curl } \mathbf{F} \text{ reduces to: } \left(\frac{\partial F_2}{\partial x} - \frac{\partial F_1}{\partial y} \right)\mathbf{k}$$

- **Example:** Consider a rotating rigid body about a fixed axis in space that is represented by vector **W** pointing in the direction of the axis of rotation, having magnitude that represents the angular speed of rotation.

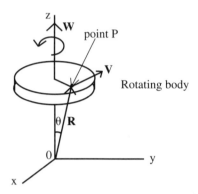

The velocity field of rotation can be represented by $\mathbf{V} = \mathbf{W} \times \mathbf{R}$, which is the velocity at point P and where $\mathbf{R}$ is the *position vector* of a point P moving with respect to a Cartesian coordinate system. If the axis of rotation is the Z-axis of the coordinate system, then:

$$\mathbf{W} = \omega_1\mathbf{i} + \omega_2\mathbf{j} + \omega_3\mathbf{k} = \omega_3\mathbf{k}$$
$$\mathbf{V}(x,y,z) = \mathbf{W} \times \mathbf{R} = \omega_3\mathbf{k} \times (x\mathbf{i} + y\mathbf{j} + z\mathbf{k})$$

where **W** points in the positive z-direction. Therefore:

$$\mathbf{V} = \begin{vmatrix} \mathbf{i} & \mathbf{j} & \mathbf{k} \\ 0 & 0 & \omega_3 \\ x & y & z \end{vmatrix} = \omega_3(-y\mathbf{i} + x\mathbf{j})$$

$$\text{curl } \mathbf{V} = \begin{vmatrix} \mathbf{i} & \mathbf{j} & \mathbf{k} \\ \partial/\partial x & \partial/\partial y & \partial/\partial z \\ -\omega_3 y & \omega_3 x & 0 \end{vmatrix} = 2\omega_3\mathbf{k} = 2\mathbf{W}$$

Therefore, curl $\mathbf{V} = 2\mathbf{W}$, and occurs in a pure rotation field. This example demonstrates that, for a rotating rigid body, the curl of the velocity field $\mathbf{V}$ has the *direction* of the axis of rotation Z and a *magnitude* equal to twice the angular speed ω of rotation. This result does not depend on the coordinate system chosen because the direction and length of curl $\mathbf{V}$ are not dependent on the choice of coordinate systems in space.

In summary, curl measures spin, the direction of curl is the axis of rotation, and the magnitude of curl is two times the speed of rotation.

• If $\mathbf{F}$ is a *gradient field*: $\mathbf{F} = (\partial f/\partial x)\mathbf{i} + (\partial f/\partial y)\mathbf{j} + (\partial f/\partial z)\mathbf{k}$, then curl $\mathbf{F} = \text{curl}(\text{grad } f) = \nabla \times \nabla f =$

$$\left(\frac{\partial}{\partial y}\frac{\partial f}{\partial z} - \frac{\partial}{\partial z}\frac{\partial f}{\partial y} \right)\mathbf{i} + \left(\frac{\partial}{\partial z}\frac{\partial f}{\partial x} - \frac{\partial}{\partial x}\frac{\partial f}{\partial z} \right)\mathbf{j} + \left(\frac{\partial}{\partial x}\frac{\partial f}{\partial y} - \frac{\partial}{\partial y}\frac{\partial f}{\partial x} \right)\mathbf{k} = 0$$

where $(\partial^2 f/\partial y\partial z) = (\partial^2 f/\partial z\partial y)$, $(\partial^2 f/\partial x\partial z) = (\partial^2 f/\partial z\partial x)$ and $(\partial^2 f/\partial y\partial x) = (\partial^2 f/\partial x\partial y)$. Because these terms cancel each other, the *curl of a gradient field* is zero: *curl(grad f)* = $\mathbf{0}$.

Because the curl characterizes the rotation in a field and the curl of a gradient field is zero, then *gradient fields* are *irrotational*.

• The *divergence of curl* $\mathbf{F}$ for every $\mathbf{F}$ is zero, because divergence represents flow away from a point and curl represents flow around a point. If:

$\mathbf{F}(x,y,z) = F_1\mathbf{i} + F_2\mathbf{j} + F_3\mathbf{k}$

div curl $\mathbf{F} = \nabla \cdot \nabla \times \mathbf{F} =$

$$\frac{\partial}{\partial x}\left(\frac{\partial F_3}{\partial y} - \frac{\partial F_2}{\partial z} \right)\mathbf{i} + \frac{\partial}{\partial y}\left(\frac{\partial F_1}{\partial z} - \frac{\partial F_3}{\partial x} \right)\mathbf{j} + \frac{\partial}{\partial z}\left(\frac{\partial F_2}{\partial x} - \frac{\partial F_1}{\partial y} \right)\mathbf{k} = 0$$

where $(\partial^2 F_1/\partial y\partial z) = (\partial^2 F_1/\partial z\partial y)$, $(\partial^2 F_2/\partial x\partial z) = (\partial^2 F_2/\partial z\partial x)$ and $(\partial^2 F_3/\partial y\partial x) = (\partial^2 F_3/\partial x\partial y)$, and these terms cancel each other.

• In general, a *rotation or spin field* has zero *divergence* and a *radial field or position vector field* $\mathbf{R}$ has zero *curl*. Note that a field with all parallel vectors may still possess rotation and, therefore, non-zero curl if the parallel vectors have different lengths that produce spin.

• Consider the following three figures representing *curl in a vector field*. The XY-plane is depicted only with the Z-axis coming out of the page.

(a.) Horizontal field:

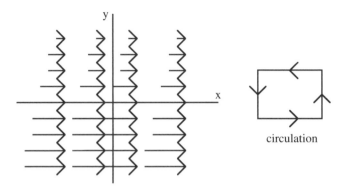

circulation

All vectors are parallel to the X-axis and point in the x-direction. However, they are of differing lengths or magnitudes and, therefore, the curl is non-zero. The sides that are perpendicular to the X-axis don't contribute to the curl or the circulation. The top vectors that are parallel to the X-axis are smaller than the bottom vectors. Therefore, the curl is non-zero and the circulation is net positive and has an upward pointing z-component by the right-hand screw rule. (See Section 5.5 for the right-hand screw rule.)

(b.) Rotational field:

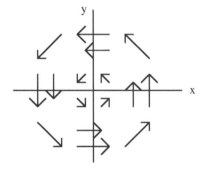

This is a rotation field and, therefore, should have a non-zero curl. Using the right-hand screw rule, the z-component of curl points upward.

(c.) Radial field:

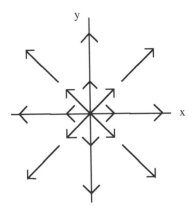

This radial field does not indicate any rotation. Therefore, the curl should be zero.

7.8 Stokes' Theorem

• This section includes the definition of Stokes' Theorem and reducing it to Green's Theorem.

• *Stokes' Theorem* transforms *line integrals* into *surface integrals* and vice versa, and also involves *curl*. Stokes' Theorem is a generalization of *Green's Theorem*, which relates line integrals to surface integrals in two dimensions.

• Stokes' Theorem states that if S is a piece-wise smooth oriented surface in space with a piece-wise smooth boundary that is a closed curve C, and $\mathbf{F}(x,y,z)$ is a continuous vector function with continuous first partial derivatives in a domain of space containing S, then the following is true:

$$\oint_C \mathbf{F} \cdot d\mathbf{R} = \iint_S (\text{curl } \mathbf{F}) \cdot \mathbf{n}dS$$

where $\mathbf{n}$ is a unit normal vector of S and integration around C has an orientation (or specified direction).

Also, $\mathbf{R} = x\mathbf{i} + y\mathbf{j} + z\mathbf{k}$ and $\mathbf{n}dS$ is often written $\mathbf{n}dA$.

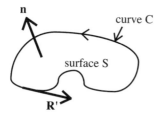

- The line integral:

$$\oint_C \mathbf{F} \cdot d\mathbf{R}$$

represents work around a curve. In Green's Theorem, the surface integral:

$$\iint ((\partial F_2/\partial x) - (\partial F_1/\partial y))\, dxdy$$

represents a surface in an XY-plane with the z-direction for **k** as normal to the surface. Stokes' Theorem involves all three components of three-dimensional space for curl **F** including the **k** component of curl.

- In Stokes' Theorem, the integral:

$$\iint_S (\text{curl } \mathbf{F}) \cdot \mathbf{n} dS$$

represents a sum of the spins or rotations in the surface and the integral:

$$\oint_C \mathbf{F} \cdot d\mathbf{R}$$

represents total circulation (or work) around curve C.

- Stokes' Theorem is reduced to Green's Theorem for a plane, where **F** = F_1**i** + F_2**j** is a vector function that is continuously differentiable in the domain of the XY-plane containing a smooth closed region S with a boundary C that is a piece-wise smooth curve. By Stokes' Theorem:

$$(\text{curl } \mathbf{F}) \cdot \mathbf{n} = (\text{curl } \mathbf{F}) \cdot \mathbf{k} = ((\partial F_2/\partial x) - (\partial F_1/\partial y))$$

where **n** is normal to the plane.

Then Stokes' Theorem becomes Green's Theorem:

$$\iint_S ((\partial F_2/\partial x) - (\partial F_1/\partial y))\, dxdy = \oint_C (F_1 dx + F_2 dy)$$

- In a gradient field, the curl is zero, therefore using Stokes' Theorem:

$$\oint_C \mathbf{F} \cdot d\mathbf{R} = \iint_S (\text{curl } \mathbf{F}) \cdot \mathbf{n}dS$$

Then:

$$\text{curl } \mathbf{F} = 0 \text{ and } \oint_C \mathbf{F} \cdot d\mathbf{R} = 0$$

Because a gradient field has zero-curl, it does no work. Remember that gradient fields are conservative fields.

Chapter

8

Introduction to Differential Equations

This chapter is designed to provide a brief classification of common or standard forms of elementary differential equations for the purpose of introducing the subject. In general, differential equations can be classified according to a few major categories. These include linear differential equations, non-linear differential equations, and systems of both linear and non-linear equations. Linear differential equations are usually easier to solve using general methods. Non-linear differential equations are more difficult to solve and often involve approximations and numerical methods. Differential equations are also classified according to the highest order of the derivative in the equation, such as first-order or second-order for equations containing a first derivative or second derivative.

8.1 First-Order Differential Equations

• This section includes a list of first-order differential equations and their general solution forms. These include simple differential equations that depend only on x, differential equations that have a real constant coefficient, initial value problems, separable equations, exact equations, linear first-order differential equations, and non-linear equations.

• *First-order differential equations* are equations representing a function that involves the first derivative of the function. Applications of first-order differential equations include modeling, electric circuits, radioactive decay, compound interest, mixing, epidemics, and elementary mechanics.

• First-order differential equations are written in the following forms:

$$F(x,y,y') = 0$$

$$y' = f(x,y)$$

$$y' + p(x)y = r(x)$$

where p and r are given continuous functions. The solution and unknown function is y, and with its derivative y' satisfies this differential equation.

• First-order differential equations can have a *general solution* that can involve a constant c and represent a *family of solutions*. Similar to indefinite integrals, the general solution of a differential equation can represent a family of curves. Similar to definite integrals, a *particular solution* of a differential equation can represent one of the curves. A particular solution of a differential equation satisfies a *specified condition*, which may be an *initial condition*.

• Following is a list of standard *differential equations* and their *solution forms*.

(a.) Equations in the form: $y' = f(x)$ are simple differential equations that depend only on x. A solution to this type of equation has the form:

$$y = \int f(t)dt + c$$

(b.) Equations in the form: $y' + ay = 0$ are differential equations that have a real *constant coefficient* a.

A solution to this type of equation can be found by inspection. A function y must be found whose derivative y' is equal to $(-a)(y)$. The solution has the form:

$y = ce^{-ax}$ where c is an arbitrary constant.

This solution represents a *family of infinitely many solutions to the differential equation*, which forms a family of *integral curves*.

(c.) Equations in the form: $y' = f(x,y)$ and $y(x_0) = y_0$ or $y' = f(x,y)$ and $y = y_0$ at $x = x_0$ are called *initial value problems*.

In these equations, x_0 and y_0 represent values of the initial condition. The *initial condition* $y(x_0) = y_0$ is used to solve for what is called a *particular solution* of the differential equation. A particular solution is the *general solution* with c specified by the initial condition.

In many applications, differential equations describe or represent a physical system or represent a mathematical model of a system where a *specified condition* must be satisfied by the solution that is inherent in the system. If this condition is an *initial condition*, such as at time $= 0$ or position $=$ point (x_0, y_0), this becomes a problem called an *initial value problem*.

Initial value problems are more specifically represented in the form:

$y' + p(x)y = r(x)$ and $y(x_0) = y_0$

where $p(x)$ and $r(x)$ are continuous functions on an open interval containing $x = x_0$. A unique function y exists that satisfies this equation and its initial condition $y(x_0) = y_0$.

(d.) Equations in the forms: $M(x)dx = -N(y)dy$, $g(y)dy = f(x)dx$ and $(dy/dx) = f(x)g(y)$ are called *separable equations*.

Separable equations can be solved by integrating each side separately. For example, a separable equation in the form: $(dy/dx) = f(x)g(y)$ can be rearranged as:

$(dy/g(y)) = f(x)dx$

and solved by integrating:

$\int (dy/g(y)) = \int f(x)dx + c$

Substitutions can sometimes be used to modify differential equations into a separable form.

(e.) Equations in the form: $M(x,y)dx + N(x,y)dy = 0$ are called *exact equations*.

In these equations, $M = u/x$ and $N = u/y$ and therefore, $u = Mx = Ny$.

Integration can occur as:

$\quad u = \int M\, \partial x + f(y)$ and $u = \int N\, \partial y + g(x)$

where $f(y)$ and $g(x)$ represent constants of integration.

In general, an exact equation $M(x,y)dx + N(x,y)dy = 0$ is one where $(M\, dx + N\, dy)$ is an exact differential such that:

$\quad du = (\partial u/\partial x)dx + (\partial u/\partial y)dy$

which yields an implicit solution $u(x,y) = c$.

Equations that are not inherently exact can be modified to an *exact form* by multiplying the non-exact equation with a function called an *integrating factor*. An integrating factor is a function that is multiplied to a differential equation to put it into a solvable form.

(f.) Equations in the form: $dy/dx + p(x)y = r(x)$ are called *linear first-order differential equations*.

A general solution for linear first-order differential equations can be developed as follows:

Integrate $dy/dx + p(x)y = r(x)$ by transforming it using $y = u(x)z(x)$ so that:

$\quad dy/dx = u(dz/dx) + z(du/dx)$

Substitute into the differential equation:

$\quad u(dz/dx) + z(du/dx) + p(x)u(x)z(x) = r(x)$

$\quad u(dz/dx) + z[(du/dx) + p(x)u(x)] = r(x)$

First consider the term $z[(du/dx) + p(x)u(x)]$ to find u:

$\quad (du/dx) + p(x)u(x) = 0$

Rearrange:

$du/u = -p \, dx$

Integrate:

$\log u = -\int p \, dx$

$u = \exp\{-\int p \, dx\}$

Substitute back into differential equation
$u(dz/dx) + z[(du/dx) + p(x)u(x)] = r(x)$:

$\exp\{-\int p \, dx\}(dz/dx) + z[(-pe^{-\int p \, dx}) + pe^{-\int p \, dx}] = r(x)$

$\exp\{-\int p \, dx\}(dz/dx) = r(x)$

Rearrange:

$dz = r(x) \, e^{-\int pdx} \, dx$

$z = \int r(x) \, e^{-\int pdx} \, dx + c$

Therefore:

$y = uz = [e^{-\int pdx}][\int r(x) \, e^{-\int pdx} \, dx + c]$

This is *the general solution of a linear first-order differential equation.*

(g.) Equations in the form: $y' = f(x,y)$ and $y(x_0) = y_0$ with a non-linear term(s) are called *non-linear differential equations.*

A general formula does not exist to solve this type of equation. However, approximate solutions and numerical solutions can be applied. For first-order linear equations, a family of solutions can exist that depends on the specification of the arbitrary constant. Whereas for non-linear equations, even though a solution containing an arbitrary constant may exist, there may be other solutions that cannot be obtained by specifying values for the constant.

Approximating solutions for differential equations includes using *direction fields*, which involve drawing or sketching families of solution curves using the slope y'. Also, approximations are made using *iteration methods*, such as *Picard's iteration* method, which is applied to initial value problems.

Non-linear differential equations can sometimes be changed into linear form by substitution of the dependent variable and solved as linear equations. The *Bernoulli equation*
$y' + p(x)y = g(x)y^n$ is an example of such an equation.

(h.) Equations in the form: $dy/dx = f(x,y)$ are sometimes called *homogeneous* when the function f does not depend on x and y separately, but only on their ratio y/x or x/y. A so-called homogeneous equation can be written in the form: $dy/dx = F(y/x)$

8.2 Second-Order Linear Differential Equations

• This section includes second-order linear differential equations, homogeneous second-order linear equations with general and particular solutions, homogeneous equations with constant coefficients, and non-homogeneous linear differential equations.

• Applications of second-order linear differential equations occur in mechanics and electrical engineering, including vibrations and resonance, mechanical vibrations, free vibrations, forced vibrations, and electrical networks.

• A *second-order differential equation* has the general form:

$$F(x,y,y',y'') = 0$$

Within this general form are equations that can be solved for y'':

$$y'' = f(x,y,y')$$

More specifically, *second-order equations* can be written in the following forms:

$$G(x)(d^2y/dx^2) + P(x)(dy/dx) + Q(x)y = R(x)$$
$$(d^2y/dx^2) + p(x)(dy/dx) + q(x)y = r(x)$$

where G, P, Q, R, p, q, r are given functions.

• A *solution* to a second-order linear differential equation on an open interval $a < x < b$ is a function $y = h(x)$ that has derivatives $y' = h'(x)$ and $y'' = h''(x)$, and satisfies the differential equation for all values of x in the interval.

- A second-order linear equation written in the form:

$$y'' + p(x)y' + q(x)y = 0$$

is called a *homogeneous second-order linear equation.*

This type of equation has a *linear combination* of solutions referred to as the *superposition or linearity principle.* Two linear independent solutions for this equation are:

$$y = y_1(x) \text{ and } y = y_2(x)$$

And they form the solutions where:

$$y = c_1y_1 + c_2y_2$$

This linear combination $(c_1y_1 + c_2y_2)$ with c_1 and c_2 as arbitrary constants provide the form of a *general solution.*

When values for c_1 and c_2 are specified as *initial conditions*, then a *particular solution* results. For example, given initial conditions $y(x_0)$ and $y'(x_0)$ where x_0 is a point within a defined interval, then c_1 and c_2 are specified so that:

$$y(x_0) = c_1y_1(x_0) + c_2y_2(x_0)$$
$$y'(x_0) = c_1y_1'(x_0) + c_2y_2'(x_0)$$

where this system has a unique solution for c_1 and c_2 if:

$$\begin{vmatrix} y_1(x_0) & y_2(x_0) \\ y_1'(x_0) & y_2'(x_0) \end{vmatrix} \neq 0$$

Therefore, when $p(x)$ and $q(x)$ are continuous on an open interval and x_0 is in the interval, then a general solution exists in the interval. More specifically, when an initial condition is specified, a particular or unique solution exists.

- Equations in the form: $ay'' + by' + cy = 0$ are called *homogeneous equations with constant coefficients.*

To solve this type of equation substitute: $y = e^{rx}$

$$a(e^{rx})'' + b(e^{rx})' + c(e^{rx}) = 0$$

Differentiate:

$$e^{rx}(ar^2 + br + c) = 0$$

where r is a *root* of the *quadratic equation*:

$$r_1 = \frac{-b + \sqrt{b^2 - 4ac}}{2a}, r_2 = \frac{-b - \sqrt{b^2 - 4ac}}{2a}$$

When $b^2 - 4ac > 0$, the general solution of the equation is:

$$y = c_1 e^{r1x} + c_2 e^{r2x}$$

When $b^2 - 4ac = 0, r_1 = r_2 = (-b/2a)$, the general solution of the equation is:

$$y = c_1 e^{r1x} + c_2 e^{r1x} = c_1 e^{-bx/2a} + c_2 e^{-bx/2a}$$

When the roots are complex, $r_1 = \lambda + i\omega$ and $r_2 = \lambda - i\omega$, the general solution of the equation is:

$$y = c_1 e^{(\lambda + i\omega)x} + c_2 e^{(\lambda - i\omega)x} = c_1 e^{\lambda x} \cos \omega x + c_2 e^{\lambda x} \sin \omega x$$

• Equations in the form: $y'' + p(x)y' + q(x)y = r(x)$ are called *non-homogeneous linear differential equations*.

In these equations, $r(x) \neq 0$ and p, q, and r are continuous on a specified interval.

A general solution to this type of equation has the form:

$$y = y_h + y_p$$

where y_h is a *general solution* of the *homogeneous equation*:

$$y'' + p(x)y' + q(x)y = 0$$

and y_p is a *particular solution of the non-homogeneous equation*:

$$y'' + p(x)y' + q(x)y = r(x)$$

Therefore, the general solution of a non-homogeneous equation combines the solution of the homogeneous equation with the particular solution y_p:

$$y = y_h + y_p = c_1 y_1 + c_2 y_2 + y_p$$

Methods used to find y_p include the method of *variation of parameters* and the method of *undetermined coefficients*.

Numerical methods and series methods are commonly used to solve second-order differential equations that have variable coefficients.

8.3 Higher-Order Linear Differential Equations

• This section includes n_{th}-order linear differential equations, n_{th}-order homogeneous linear differential equations, n_{th}-order homogeneous equations with constant coefficients, n_{th}-order non-homogeneous linear differential equations, and n_{th}-order non-homogeneous differential equations with constant coefficients.

• Higher-order linear differential equations are an extension of second-order linear differential equations as far as form and solution methods.

• An n_{th}-order linear differential equation has the general form:

$$P_0(x)(d^{(n)}/dx^{(n)}) + P_1(x)(d^{(n-1)}/dx^{(n-1)}) + \dots + P_{n-1}(x)(dy/dx)$$
$$+ P_n(x)y = r(x)$$

Or equivalently:

$$P_0(x)y^{(n)} + P_1(x)y^{(n-1)} + \dots + P_{n-1}(x)y' + P_n(x)y = r(x)$$

where r and P_n are continuous in a specified interval.

If the equation is divided by $P_0(x)$ it becomes:

$$y^{(n)} + p_1(x)y^{(n-1)} + \dots + p_{n-1}(x)y' + p_n(x)y = r(x)$$

• The standard form of an *nth-order homogeneous differential equation* is:

$$y^{(n)} + p_1(x)y^{(n-1)} + \dots + p_{n-1}(x)y' + p_n(x)y = 0$$

where $y^{(n)} = d^n y/dx^n$ is the first term.

• For n_{th}-*order homogeneous linear differential equations* in the form:
$y^{(n)} + p_1(x)y^{(n-1)} + \dots + p_{n-1}(x)y' + p_n(x)y = 0$,
linear combinations of solutions form a solution, (similar to second-order equations). This is called a *basis* of solutions and is comprised of *n linearly independent solutions*.

The *general solution to n_{th}-order homogeneous linear differential equation* is the *linear combination*:

$$y = c_1y_1 + c_2y_2 + \dots + c_ny_n$$

where $c_1,...c_n$ are constants and $c_1, ..., c_n$ satisfy:

$$c_1y_1(x_0) + ... + c_ny_n(x_0) = y_0$$
$$c_1y_1'(x_0) + ... + c_ny_n'(x_0) = y_0'$$
$$\vdots$$
$$c_1y_1^{(n-1)}(x_0) + ... + c_ny_n^{(n-1)}(x_0) = y_0^{(n-1)}$$

When values are specified for $c_1,...c_n$, a *particular solution* results. To obtain a unique solution, it is necessary to specify n *initial conditions*:

$$y(x_0) = y_0, y'(x_0) = y_0', ..., y^{(n-1)}(x_0) = y_0^{(n-1)}$$

In general, when $p_0,...p_{(n-1)}$ are continuous on an open interval and x_0 is in that interval, then a general solution can be obtained. If initial conditions are given, then a particular solution can be obtained.

• Equations in the form: $a_0y^n + a_1y^{(n-1)} + ...+ a_{n-1}y' + a_ny = 0$ are called n_{th}-*order homogeneous equations with constant coefficients*.

Solving this type of equation is similar to solving second-order homogeneous equations with constant coefficients. A solution involving $y = e^{rx}$ can be found.

Substituting $y = e^{rx}$ into the equation gives:

$$e^{rx}(a_0y^n + a_1y^{(n-1)} + ... + a_{n-1}y' + a_ny) = 0$$

When *roots* r are real and unequal, the *general solution* is:

$$y = c_1e^{r_1x} + c_2e^{r_2x} + ... c_ne^{r_nx}$$

When the roots are complex, $r_1 = \lambda + i\omega$ and $r_2 = \lambda - i\omega$, the general solution is:

$$y = c_1e^{\lambda x} \cos \omega x + c_2e^{\lambda x} \sin \omega x + ...$$

• Equations in the form:

$$y^{(n)} + p_1(x)y^{(n-1)} + ... + p_{n-1}(x)y' + p_n(x)y = r(x)$$

with $r(x)$ continuous on the open interval, are called *non-homogeneous* n_{th}-*order linear differential equations*.

A general solution exists in the form:

$$y = y_h + y_p = c_1y_1 + c_2y_2 + ... + c_ny_n + y_p$$

where y_h is a *general solution* of the *homogeneous equation* and y_p is added as the *particular solution* of the non-homogeneous equation.

- Equations in the form:

$$y^{(n)} + a_1 y^{(n-1)} + \dots + a_{n-1} y' + a_n y = r(x)$$

are called *non-homogeneous n_{th}-order equations with constant coefficients*.

To solve this type of equation, the *method of undetermined coefficients* and the *method of variation of parameters* can be used. Methods used for constant coefficients often involve sine, cosine, and exponential functions.

- If the coefficients are not constants, solutions often involve numerical methods or series methods. In general, methods used for solving second-order differential equations can often be expanded to higher order differential equations.

8.4 Series Solutions to Differential Equations

- This section briefly describes series solutions for differential equations with variable coefficients, the power series method, and the Frobenius method.

- Series solutions can be applied to solve linear differential equations that have *variable coefficients*.

- *Differential equations with variable coefficients* can arise in modeling applications and can be in the general form:

$$P(x)y'' + Q(x)y' + G(x)y = 0$$

$$\text{or } y'' + p(x)y' + q(x)y = 0$$

where the coefficients P, Q, G, p, and q are polynomials.

- *Series solution methods* for a differential equation with variable coefficients involve solving the equation near a point x_0. Using a series solution method generally involves expressing y as an *infinite series in powers* of $(x - x_0)$, where x_0 is a specified point.

- The *power series method* is a general method for solving linear differential equations in the form:

$$y'' + p(x)y' + q(x)y = r(x)$$

(including higher orders) where $p(x)$, $q(x)$, and $r(x)$ are variable.

The power series method provides solutions in the form of the power series:

$$y(x) = a_0 + a_1(x - x_0) + a_2(x - x_0)^2 + \ldots$$

In this method, the power series is substituted along with its derivatives into the differential equation:

$$y'' + p(x)y' + q(x)y = r(x)$$

The coefficient a_n can therefore be determined, providing p, q, and r are analytic at $x = x_0$.

Note: A function $f(x)$ is said to be *analytic* if it is differentiable at all points in its domain. A function $f(x_0)$ is analytic if it is differentiable at and near point x_0. Also, a function that is real and analytic at point $x = x_0$ can be represented in a power series in powers of $(x - x_0)$ with a positive radius of convergence.

- The *Frobenius method* allows the power series to be extended to differential equations in the form:

$$y'' + [b(x)/(x - x_0)]y' + [c(x)/(x - x_0)^2]y = 0$$

where the coefficients are singular (cannot be obtained from a general solution) at $(x = x_0)$ rather than analytic, however $b(x)$ and $c(x)$ are analytic at $(x = x_0)$.

These equations can have a solution in the form:

$$y(x) = x^r[a_0 + a_1(x - x_0) + a_2(x - x_0)^2 + \ldots]$$

where r is a real or complex number that is determined by substituting $y(x)$ into the differential equation.

8.5 Systems of Differential Equations

• This section provides a brief introduction to systems of linear differential equations, including systems of first-order differential equations, systems of linear differential equations with constant coefficients a_{ij}, and systems of homogeneous linear differential equations with constant coefficients a_{ij}.

• *Systems of differential equations* include linear systems and non-linear systems. Systems of linear differential equations can also be homogeneous or non-homogeneous, and can be solved using methods that include vectors and matrices and phase-plane methods. Systems of higher-order differential equations can sometimes be reduced to first-order equations so that simpler methods can be applied to solve them.

• Applications of systems of differential equations include mechanical systems containing springs or masses, combined networks of circuits, and many other systems in various disciplines of engineering.

• In general, a *system of first-order differential equations* has the form:

$$y_1' = f_1(t, y_1, y_2, y_3)$$
$$y_2' = f_2(t, y_1, y_2, y_3)$$
$$y_3' = f_3(t, y_1, y_2, y_3)$$

or in more general form:

$$y_1' = f_1(t, y_1, ..., y_n)$$
$$y_2' = f_2(t, y_1, ..., y_n)$$
$$\vdots$$
$$y_n' = f_n(t, y_1, ..., y_n)$$

In such a system of differential equations, the unknown functions in the equations are solved.

• A system of differential equations in the form:

$$y_1' = a_{11}y_1 + a_{12}y_2 + g_1$$
$$y_2' = a_{21}y_1 + a_{22}y_2 + g_2$$

is a *linear system of differential equations* with *constant coefficients* a_{ij}.

This system can also be written in *vector* form as:

$$\mathbf{y'} = \mathbf{Ay} + \mathbf{g}$$

where $\mathbf{A} = \begin{bmatrix} a_{11} & a_{12} \\ a_{21} & a_{22} \end{bmatrix}, \mathbf{y} = \begin{bmatrix} y_1 \\ y_2 \end{bmatrix}, \mathbf{g} = \begin{bmatrix} g_1 \\ g_2 \end{bmatrix}$

• If this system of *linear differential equations* with *constant coefficients* a_{ij} has $\mathbf{g} = 0$, then it becomes a *homogeneous linear system of differential equations* with constant coefficients a_{ij} and can be written:

$$y_1' = a_{11}y_1 + a_{12}y_2$$
$$y_2' = a_{21}y_1 + a_{22}y_2$$

In vector form, these equations become:

$$\mathbf{y'} = \mathbf{Ay}$$

where $\mathbf{A} = \begin{bmatrix} a_{11} & a_{12} \\ a_{21} & a_{22} \end{bmatrix}, \mathbf{y} = \begin{bmatrix} y_1 \\ y_2 \end{bmatrix}$

Solutions to a system of homogeneous linear equations have the form:

$$\mathbf{y} = \mathbf{x}e^{\lambda t}$$

where λ is an *eigenvalue* of $\mathbf{A}$ and $\mathbf{x}$ is the *eigenvector*. The solution to the quadratic equations represented below is λ:

$$\begin{bmatrix} a_{11} - \lambda & a_{12} \\ a_{21} & a_{22} - \lambda \end{bmatrix} = (a_{11} - \lambda)(a_{22} - \lambda) - a_{12}a_{21} = 0$$

where eigenvector $\mathbf{x} \neq 0$, and together with its components x_1 and x_2 form:

$$(a_{11} - \lambda)x_1 + a_{12}x_2 = 0$$
$$a_{12}x_1 + (a_{22} - \lambda)x_2 = 0$$

• Note that a system of differential equations can be solved using a *phase-plane method* where solutions to:

$$\mathbf{y'} = \mathbf{Ay}$$

or equivalently, for two dimensions:

$$y_1' = a_{11}y_1 + a_{12}y_2$$
$$y_2' = a_{21}y_1 + a_{22}y_2$$

are found such that $y_1 = y_1(t)$ and $y_2 = y_2(t)$ exist as a path or curve of a solution in a y_1y_2-phase plane. A point $P(y_1,y_2)$ is a critical point of the system and occurs where the right sides of the system equal zero. Point P can be a node, saddle point, center, or spiral point, and can be stable or unstable. (Please see a textbook on differential equations for a complete explanation of this and other solutions in this chapter.) Phase-plane methods can be applied to non-linear systems using linearization.

8.6 Laplace Transform Method

- This section provides a brief introduction to the Laplace transform method for solving differential equations.

- The *Laplace transform method* is used to solve differential equations and systems of differential equations and their corresponding initial and boundary value problems. The method involves transforming a complicated problem into a simple equation called a *subsidiary equation*, solving this equation using algebraic techniques, then transforming the solution of the subsidiary equation back to find the solution of the original problem.

- The *Laplace transform* of a function f(t) is written:

$$F(s) = \mathcal{L}(f) = {}_0\!\int^{\infty} e^{-st} f(t) \, dt$$

where differentiation of f with respect to t corresponds to the multiplication of the transform F with s:

$$\mathcal{L}(f'(t)) = s\mathcal{L}(f(t)) - f(0)$$

$$\mathcal{L}(f''(t)) = s^2\mathcal{L}(f(t)) - sf(0) - f'(0)$$

$$\mathcal{L}(f^{(n)}(t)) = s^n\mathcal{L}(f(t)) - s^{(n-1)}f(0) - \ldots - sf^{(n-2)}(0) - f^{(n-1)}(0)$$

- To solve a given differential equation in the form:

$$y'' + ay' + by = r(x)$$

First take the transform and set $\mathcal{L}(y) = Y(s)$ to determine a *subsidiary equation* that has the form:

$$(s^2 + as + b)Y = \mathcal{L}(r) + sf(0) + f'(0) + af(0)$$

Tables of functions and their Laplace transforms $\mathcal{L}(f)$ can be used to obtain the transform $\mathcal{L}(r)$.

The subsidiary equation is solved for $Y(s)$ algebraically and the inverse transform $y(t) = \mathcal{L}^{-1}(Y)$ is determined to find the solution. This last step often involves using Laplace transform tables.

8.7 Numerical Methods for Solving Differential Equations

- This section provides a brief introduction to the use of numerical methods for solving differential equations including the Euler method, the Improved Euler method, the Runge-Kutta method, and the Adams-Moultan method.

- *Numerical methods* are used to solve various types of differential equations. Numerical procedures involve constructing approximate values of $y_0, y_1, y_2,...y_n$ at points $x_0, x_1, x_2,...x_n$. Problems to consider when using numerical methods include convergence and error.

- To demonstrate the concept of numerical methods, consider a first-order initial value problem:

$$y' = f(x,y), y(x_0) = y_0$$

To find the solution, begin with the *Taylor series*:

$$y(x+h) = y(x) + hy'(x) + (h^2/2)y''(x) +...$$

Then truncate the series after the y' term. This results in an expression used repeatedly in the *Euler method*:

$$y_{n+1} = y_n + hf(x_n,y_n) = y_n + hy_n'$$

where $n = 0, 1, 2, ...,$ and h is the step size between points $x_0, x_1, x_2,...x_n$.

If the series is truncated to include the y″ term, then the resulting expression is used in the *Improved Euler method* (also called the *Improved Euler-Cauchy method* or *Heun's method*):

$$y_{n+1} = y_n + h[f(x_n,y_n) + f(x_n+h, y_n+hf(x_n,y_n))]/2$$
$$= y_n + h[y_n' + f(x_n+h, y_n+h\ y_n')]/2$$

If the series is truncated to include the h⁴ term, a more accurate method results called the *Runge-Kutta method of fourth-order*. This method involves calculating the following:

$$k_{n1} = f(x_n, y_n),$$
$$k_{n2} = f(x_n + h/2, y_n + hk_{n1}/2)$$
$$k_{n3} = f(x_n + h/2, y_n + hk_{n2}/2)$$
$$k_{n4} = f(x_n + h, y_n + hk_{n3})$$

Then substituting them into the expression:

$$y_{n+1} = y_n + (h/6)[k_{n1} + 2k_{n2} + 2k_{n3} + k_{n4}]$$

• Another numerical method called the *Adams-Moultan method* involves calculating a "predictor" given by:

$$y_{n+1} = y_n + (h/24)[55y_n' - 59y_{n-1}' + 37y_{n-2}' - 9y_{n-3}']$$

Then calculating a "corrector" given by:

$$y_{n+1} = y_n + (h/24)[9y_{n+1}' + 19y_n' - 5y_{n-1}' + y_{n-2}']$$

where y_1, y_2, y_3 are first calculated using the Runge-Kutta method.

• Second-order ordinary differential equations can be solved using an extension of the Runge-Kutta method called the Runge-Kutta-Nystrom method.

• Note that numerical methods are commonly used to solve partial differential equations.

8.8 Partial Differential Equations

• This section provides a brief introduction to partial differential equations.

• *Partial differential equations* are used to model physical and geometrical systems where there are functions that depend on two or more independent variables. Partial differential equations arise in fluid mechanics, dynamics, elasticity, heat transfer, quantum mechanics, electro-magnetic theory, and many engineering problems. In partial differential equations, the independent variables include time and space coordinates.

• Examples of second-order partial differential equations include:

One-dimensional wave equation:

$$\partial^2 u/\partial t^2 = c^2(\partial^2 u/\partial x^2)$$

Two-dimensional wave equation:

$$\partial^2 u/\partial t^2 = c^2[(\partial^2 u/\partial x^2) + (\partial^2 u/\partial y^2)]$$

One-dimensional heat equation:

$$\partial u/\partial t = c^2(\partial^2 u/\partial x^2)$$

Two-dimensional Laplace equation:

$$\nabla^2 u = (\partial^2 u/\partial x^2) + (\partial^2 u/\partial y^2) = 0$$

Three-dimensional Laplace equation:

$$\nabla^2 u = (\partial^2 u/\partial x^2) + (\partial^2 u/\partial y^2) + (\partial^2 u/\partial z^2) = 0$$

Two-dimensional Poisson equation:

$$(\partial^2 u/\partial x^2) + (\partial^2 u/\partial y^2) = f(x,y)$$

• Solutions to partial differential equations are often obtained in a specified region that satisfies *initial conditions* or *boundary conditions* where values of the solution u or its derivatives on the boundary curve or surface of the region are set. For example, in wave equations initial conditions may be displacement or velocity at time $t = 0$. Or in heat equations an initial temperature may be specified.

• Partial differential equations can be solved using a *separation of variables* method or the *product method* in which the solutions form products of functions that each depend on one of the variables. For example, the solution form $u(x,t) = F(x)G(x)$ can be used to solve the one-dimensional wave equation or the one-dimensional heat equation, where substituting into the partial differential equation gives an ordinary differential equation for F and G.

• *Numerical methods* are commonly used to solve partial differential equations. Such methods can include replacing the partial derivatives with *difference quotients*. The following solution forms can also be used to solve these equations:

For the Laplace equation:

$$u_{i+1,j} + u_{i,j+1} + u_{i-1,j} + u_{i,j-1} - 4u_{i,j} = 0$$

For the heat equation:

$$(1/k)[u_{i,j+1} - u_{i,j}] = (1/h^2)[u_{i+1,j} - 2u_{i,j} + u_{i-1,j}]$$

For the wave equation:

$$(1/k^2)[u_{i,j+1} - 2u_{i,j} + u_{i,j-1}] = (1/h^2)[u_{i+1,j} - 2u_{i,j} + u_{i-1,j}]$$

where h and k represent the size of the sections in a grid in x- and y-directions.

Index

A

a^x, derivatives of, 71–77
acceleration, 54, 81, 230–241
  centripetal, 22
  curves, 239
  integrals, 119
  vectors, 236, 241
addition
  of complex numbers, 37
  formulas, 10, 14
  formulas, for cosine, 92
  formulas, for sine, 92
  of functions, 5
  of matrices, 205
  of vectors, 199–202
algebra, matrices and linear, 217–224
alternating series, 186
angles
  double angle formula, 15
  measurements, 11
  polar, 36
  rate of revolutions, 21
  between vectors, 210
  velocity, 21
antiderivatives, 117–120
applications
  of exponential equations, 77–80
  integrals, 157–161
approximating
areas, 127
  linear approximations, 250–254
  quadratic approximations, 250–254
  slopes of curves, 109–112
arcs
  of cycloid, 234
  length of, 10–11
Arcsine, 19
Arctangent, 19
areas
  approximating, 127

below X-axis, 131–134
  under curves, 124–127
  of cycloid, 235
  derivatives, 60–63
  integrals, 140–145
  of parallelograms, 213, 291
  rectangles, 125
arithmetic
  progressions, 180
  series, 181
arrays, 203
associations, 3
  errors, sums and, 128–131
asymptotes, 45
averages
  flow, 299
  slopes, 57
  values, integrals, 131
  velocity, 54, 57
axes, 148

B

base 10 logarithms, 9
base numbers, changing from one to the other, 10
binomial expressions, 194
Binomial Theorem, 194
boundary conditions, 324

C

calculating. *See also* formulas
  derivatives, 72
  determinants, 157, 212
  functions, 109
  integrals, 121
Cartesian coordinates, 33, 228, 296
Cauchy-Riemann equations, 287
centers
  of mass, 159–161
  of the series, 188
centripetal acceleration, 22
chain rule, 69

partial derivatives, 246–247
changing
  coordinates, 152–157
  variables, 116, 152–157
circles, 10
  areas of, 140
  circular functions, 10
  circular motion, 20–25
  convergence, 193
  on coordinate systems, 35
  distance between points, 13
  ellipses and, 42–44
  length of, 229
  motion of particles in, 231
  particles moving around, 24
  radius. *See also* radius
  trigonometric functions, 11
circumference, 140, 229.
  See also circles
closed curves, 282
closeness of functions, 47
coefficients
  constants, 308
  matrices of, 203, 216
columns, 147
  multiplication, 207
  vectors, 196, 220
common differences, 180
common integrals, 138–139
common ratios, 181
comparing integrals, 136
Comparison Tests, 123–124, 184–185
complex numbers, 36–38
complex planes, 38
components, 196–197
composite functions, 6
  chain rule, 87
compound functions, 6
concave-down shapes, 94
concave-up shapes, 94

conditions, 49, 308
for infinite series, 183
initial, 324
conservative fields, 274
vectors, 276–282
constants, 8, 117
coefficients, 308
derivatives, 63–64
functions, 66–67
integrals, 134–137
integration, 118, 121
variables, 66–67
velocity, 61
constrained optimization, 263–265
continuable functions, 50
continuity, 50–52
continuous functions, 50–51
contour diagrams, 29–33, 274
convergence, 49, 122
circles, 193
intervals, 189, 192
radius, 189
tests, 183–184
coordinate systems, 33–36
changing, 152–157
cylindrical, 149
polar, 156
rectangles, 224
spheres, 227
trigonometric functions, 10
vectors, 197
cosecant, 10, 17
derivatives, 95
hyperbolic, 28
cosine, 7, 10, 17, 51
addition formulas for, 14
curves, 24, 92
directions, 199
hyperbolic, 26
inverse, 98
cotangents, 10, 17
derivatives, 95
hyperbolic, 28
Cramer's rule, 126, 215, 220
critical points, 261
cross products, 211–215
cubes, 148
volume, 215
curl, 300–304
curves, 8, 92, 224–229
acceleration, 239
areas under, 124–127
closed, 282
continuity, 50
cosine, 24
flux, 288
length of, 139
level, 31
moving, 232

orientation of, 277
sections of, 140
sine, 24
slopes, approximating, 109–112
cycloid, 233–234
cylinders, 226
coordinates, 34, 149
shells, volume of, 146

D

decay models, 77–80
definite integrals, 120–122, 127, 163, 277
definitions
of derivatives, 53–56, 72
of determinants, 215–217
of functions, 3–8
of limits, 47–50
of matrices, 202–205
partial derivatives, 244
of rate of change, 53–56
of relationships, 25–26
of vectors, 195–199
degrees, 10
polynomial functions, 28
delta (Δ)notation, 57–58
product rules, 83
denominators, 28
dependent variables, 29, 243
derivatives, 53
of a^x, e^x, and x, 71–77
antiderivatives, 117–120
areas, 60–63
chain rule, 86–90
constants, 63–64
cosecant, 95
cotangents, 95
definitions of, 53–56, 72
delta (Δ) notation, 57–58
differentiating, 68–69
directional, 268–270
directional, gradients and, 255–259
distance, 60–63
first, 103–109
formulas, 64–66
functions, 69–71
hyperbolic functions, 99–100
inverse functions, 95–99
maximum, 103–109
minimum, 103–109
multiple, 82
of multiple products, 85
of natural logarithms, 76
negatives, 94
partial. See partial derivatives
points, 59

positive, 94
powers, 69–71
of products, 82–85
of quotients, 85–86
rate problem examples, 90–91
reciprocal functions, 89
secant, 95
second, 81–82, 103–109
slopes, 60–63
tangents, 95
variables, 66–67
vectors, 237
velocity, 60–63
zeroes, 104
determinants
calculating, 157, 212
definitions, 215–217
Jacobian, 154
diagrams, contour, 29–33, 274. See also graphs
differences, 80–81
common, 180
quotients, 7, 325
differentiability, 52–53
differential equations, 250–254, 307
first-order, 308–312
high-order linear, 314–317
Laplace transform method, 321–322
numerical methods for solving, 322–323
partial, 324–325
second-order linear, 312–314
series, 317–318
systems of, 319–321
differentiating, 60
chain rule, 86–90
derivatives, 68–69
explicit functions, 101–102
hyperbolic functions, 99–100
implicit functions, 101–102
matrices, 205
multivariable functions, 101
numerical methods, 111
quotients, 95
rules, 102–103
series, 189
sums, 80–81
trigonometric functions, 91–95
directional derivatives, 268–270
and gradients, 255–259
directions, 195

cosines, 199
of lines, 230
motion, 238
vectors, 199
discontinuous functions, 51
discontinuous integrands, 123
displacement vectors, 196,
210, 235, 279
distance, 210
center of mass, 159–161
derivatives, 60–63
horizontal, 233
integrals, 124
between points, 12
between points, circles, 13
between two points, 10
distributions, 161
divergence, 49, 183, 295–300
series, 185
division
of complex numbers, 37
of functions, 5
domains, 180
sets, 4
trigonometric function
values, 20
dot products, 207, 208–211
gradients, 258
double angle formulas, 10, 15
double integrals, 136–138,
154
double-subscript notation,
204
dummy variables, 114

E

e^x, derivatives of, 71–77
elements
in matrices, 203
in range sets, 3
ellipses and circles, 42–44
entries, matrices, 203
equal matrices, 204
equalpotentials, 275
equations, 3
for asymptotes, 45
Cauchy-Riemann, 287
Cramer's rule, 215–216
differential. *See* differential
equations
exponential, applications
of, 77–80
Laplace's, 287
linear, 218
for line normals, 60
of lines, 230
Newton's method for, 109,
111
non-linear, 7

for parabolas, 39
parameters, 224–229, 232
planes, 226
for planes, 30
quadratic, 41
subsidiary, 321
tangent lines, 59, 109
equivalent vectors, 197
errors
associated errors, sums
and, 128–131
trapezoid rules, 129
estimating
infinite series, 183
infinite series, sums of, 49
integrals, 114, 128–131
regions, 59
Euler's formula, 193
Euler's identity, 26
Euler's method, 322
evaluating
derivatives, 63–64
integration, 162–177
partial derivatives,
243–246
even functions, 7, 27,
131–134
exact equations, 310
exact forms, 310
examples, rate problem,
90–91
expanding
series, 188–194
trigonometric functions, 38
explicit functions, differentiating, 101–102
exponential equations, applications of, 77–80
exponential functions, 3,
8–10, 51, 71, 192
relationships, 25–26
expressions, 3
binomial, 194
polynomial functions, 28
extrema
local, 259
points, 104

F

factoring polynomials, 48
families
of functions, 119
of solutions, 308
fields
conservative, 276–282
divergence, 295
gradients, 274
lines, 271–276
slopes, 119

vectors, 271–276
finite arithmetic progressions,
180
finite sequences, 179
first derivatives, 103–109
first-order differential
equations, 308–312
flow
averages, 299
fields, 275
lines, 276
flux, 287–295
foci, 43
focus points, 41
force, 159
dot products, 210
fields, 274
forms
exact, 310
solution, 308
formulas, 4
addition, 10, 14
derivatives, 64–66, 68–69
double angle, 10, 15
Euler's, 193
integrals, 117–120
products, 83
progressions, 182
Pythagorean, 10, 12
subtraction, 10, 14
fractions, evaluating, 172–177
Frobenius method, 318
functions. *See also* fields
calculating, 109
chain rule, 86–90
circular, 10
circular motion, 20–25
closeness of, 47
complex numbers, 36–38
composite, 6, 87
compound, 6
constants, 66–67
continuable, 50
continuous, 50, 51
coordinate systems,
33–36
derivatives, 69–71
discontinuous, 51
domain sets, 4
even, 7, 27, 131–134
explicit, differentiating,
101–102
exponential, 8–10, 51, 71,
192
families of, 119
graphs, 4, 70
hyperbolic, 26–28
hyperbolic, differentiating,
99–100

functions *(Continued)*
  implicit, differentiating,
    101–102
  integration, 134–136
  inverse, 6
  inverse, derivatives, 95–99
  Lagrangian, 265
  linear, 6, 63–64
  local extrema of, 104
  logarithmic, 8–10
  with multiple variables,
    29–33
  multivariable, differentiat-
    ing, 101
  non-linear, 6–7
  odd, 7, 8, 27, 131–134
  one-variable, 30, 260
  polynomial, 28, 50, 80–81
  potential, 274
  rational, 51
  reciprocal, 89
  relationships, 25–26
  slope, 6
  stream, 287
  three-variable, 30
  trigonometric, 10–20
  trigonometric, derivatives,
    95–99
  trigonometric, differentiat-
    ing, 91–95
  trigonometric, inverse, 19
  two-variable, 30, 141
  types of, 3–8
Fundamental Theorem of
  Calculus, 120–122

G

Gauss elimination, 220
Gauss-Jordan elimination,
  220
Gauss's Divergence Theorem,
  298
generalized Ratio Tests, 186
general solutions, 308
geometry
  progressions, 181
  series, 182, 184, 193
globals, 104
gradients, 268–270
  directional derivatives,
    and, 255–259
  fields, 274
graphs
  continuous functions, 51
  functions, 4, 70
  of inverse functions, 96
  limits, 47
  multiple variables, 29–30
  pairs, 5

parabolas, 41
partial derivatives,
  247–250
of trigonometric functions,
  10, 16
gravitational fields, 274
gravity, 277
Green's Theorem, 277,
  282–287
growth models, 77–80

H

harmonic motion, 20, 22
Harmonic Series, 185
height, maximum, 233
Heun's method, 323
high-order linear differential
  equations, 314–317
homogeneous systems, 218
horizontal distances, 233
horizontal fields, 272
hyperbolas, 44
hyperbolic functions, 26–28
  differentiating, 99–100
hypotenuse, 33

I

identity matrices, 219
implicit functions, 101–102
improper integrals, 122–124
Improved Euler method, 323
increments of time, 54
indefinite integrals, 117–120,
  135
independent variables, 29,
  243
infinite geometric series, 182
infinite sequences, 179
infinite series, 183–184, 188
  sums of, 49
infinity, convergence, 122
initial conditions, 309, 324
initial value problems, 309
inner products, 208
instantaneous rate of change,
  56
instantaneous velocity, 57
integrals, 113
  acceleration, 119
  applications, 157–161
  areas, 140–145
  areas under curves,
    124–127
  average values, 131
  calculating, 121
  common, 138–139
  comparing, 136
  definite, 120–122, 127,
    163, 277

double, 154
estimating, 114, 128–131
evaluating, 162–177
formulas, 117–120
improper, 122–124
indefinite, 117–120, 135
length, 139–140
lines, 276–282
multiple, 136–138
properties, 134–136
splitting, 123
sums and sigma (θ)
  notation, 114–117
surfaces, 287–295
triple, 154
volume, 145–152
Integral Test, 187
integrands, 117, 122
  discontinuous, 123
integration, 60, 161
  constants, 118, 121
  functions, 134–136
  limits of, 153
  by parts, 162
intervals, 19
  convergence, 189, 192
  X-axis, 148
inverse cosine, 98
inverse functions, 6
  derivatives, 95–99
inverse matrices, 219
inverse sine, 98
inverse trigonometric
  functions, 10, 19
iteration, Picard's iteration
  method, 311

J

Jacobian determinants, 154
Jacobian J method, 153
J factor, 155
jumps, 51

L

Lagrange multipliers, 264
Lagrangian function, 265
Laplace's equations, 287
Laplace transform method,
  321–322
length, 195–196
  of arcs, 10–11
  of cycloid, 234
  of integrals, 139–140
  of quarter circles, 229
  of vectors, 213
level curves, 31
limits
  continuity, 50–52
  definition of, 47–50

differentiability, 52–53
of integration, 153
of sums, 114, 116
switching, 135
linear algebra, matrices, 217–224
linear approximations, 250–254
linear functions, 6, 63–64
linearity principles, 313
linearizing regions, 59
linear one-variable functions, 30
linear two-variable functions, 30
lines
contours, 31
curves. *See* curves
fields, 271–276
flow, 276
integrals, 276–282
motion in, 230
normals, 60
parametric equations for, 224
secant, 60
straight, 62
tangents, 53, 58–60
tangents, equations, 109
vertical line tests, 5
local extrema, 259
of functions, 104
local linearity, 109–112, 250–254
logarithms
functions, 3, 8–10
natural, 75

M

Maclaurin series, 188–189, 191
magnitude, vectors, 195–196, 208, 237
major axis, 43
mapping
contours, 31
radius, 20
mass, center of, 159–161
matrices
of coefficients, 216
definitions, 202–205
identity, 219
inverse, 219
linear algebra, 217–224
multiplication, 205–208
transposing, 219
triangles, 220
maxima, 259–263
maximum derivatives,

103–109
maximum height, 233
measurements
of angles, 11
of Z-axis, 34
methods
of determinants, 216, 218
Euler's, 322
Frobenius, 318
Heun's, 323
Jacobian J, 153
Laplace transform, 321–322
numerical, 109–112
numerical differentiation, 111
numerical for solving differential equations, 322–323
Picard's iteration, 311
products, 324
Runge-Kutta method of fourth-order, 323
midpoint rule, 128
minima, 259–263
minimum derivatives, 103–109
models
decay, 77–80
growth, 77–80
motion, 230–241
circular, 20–25
directions, 238
harmonic, 20, 22
oscillatory, 22
polar coordinates, 239
moving. *See also* motion
curves, 232
objects, 224
particles, 235
multiple derivatives, 82
multiple integrals, 136–138
multiple products, derivatives of, 85
multiple variables, functions with, 29–33
multiplication
complex numbers, 37
of functions, 5
matrices, 205–208
scalars, 206
vectors, 205–208
multipliers, Lagrange, 264
multivariable functions, differentiating, 101

N

natural logarithms, 9, 75
derivatives of, 76

negatives
derivatives, 94
of vectors, 200
Newton's method for equations, 109, 111
non-homogeneous systems, 218
non-linear
equations, 7
functions, 6–7
normals
line, 60
vectors, 238
notation
for delta (Δ), 57–58
for delta (Δ), product rules, 83
for derivatives, 53
for directional derivatives, 255
for double-subscript, 204
of gradients, 257
for integration, 162
for line integrals, 278
for matrices, 203
for multiple derivatives, 82
for sigma (θ), 114–117
for vectors, 196
nth derivatives, 54
nth-order homogeneous differential equations, 315
nth-order homogeneous linear differential equations, 315
numbers
complex, 36–38
in range sets, 3
sequences, 114
numerical differentiation methods, 111
numerical methods, 109–112
for solving differential equations, 322–323

O

objects
moving, 224
volume of, 145
odd functions, 7, 8, 27, 131–134
one-variable functions, 30, 260
optimization, 259–263
constrained, 263–265
orientation, 277
oscillatory motion, 22
velocity, 23
outflow, 297

P

pairs, 4, 5
parabolas, 38–42
paralleledpipeds, volume of, 214
parallelograms
 areas, 291
 areas of, 213
parallel vectors, 209
parameters, 278
 equations, 224–229, 232
 flux, 294
 planes, 225
partial derivatives
 chain rule, 246–247
 evaluating, 243–246
 graphs, 247–250
 representation, 243–246
partial differential equations, 324–325
partial fractions, evaluating, 172–177
particles
 motion in planes, 230
 moving around circles, 24
 speed of, 238
 velocity, 235
parts, integration by, 162
perpendicular vectors, 209
perspective in contour diagrams, 32
Picard's iteration method, 311
planes, 30
 complex, 38
 equations, 226
 points, 225
 tangents, 261
 XY-planes, 141
plotting, spiral of Archimedes, 35
points
 critical, 261
 derivatives, 59
 discontinuous functions, 51
 distances between, 12–13
 extrema, 104
 focus, 41
 planes, 225
 saddle, 263
 slopes, 57
 trigonometric functions, 10
 in vector fields, 272
 vertex, 39
polar angles, 36
polar coordinates, 33, 156
 motion, 239
polynomial functions, 28, 50, 80–81
 factoring, 48

integrals, 134–136
position vectors, 197, 224–229, 237, 273
positive derivatives, 94
potential functions, 274
powers
 derivatives, 69–71
 second, functions raised to, 85
 series, 188, 190, 318
 variables raised to, 66–67
pressure, 157–159
probabilities, 161
products
 cross, 211–215
 derivatives of, 82–85
 dot, 207–211
 dot, gradients, 258
 inner, 208
 methods, 324
 representations of, 117
 rules, 162
 scalars. See scalars
 vectors, 211–215
progressions, 179–182
projectiles, 232
properties, 3
 of addition, matrices and vectors, 205
 of cross products, 215
 of dot products, 211
 of hyperbolic cosines, 27
 of integrals, 134–136
 of line integrals, 279
 of multiplication of matrices, 205
 of sums, 115
Pythagorean formulas, 10, 12
Pythagorean Theorem, 33

Q

quadratic approximations, 250–254
quadratic equations, 41
quarter circles, length of, 229
quotients
 derivatives, 85–86
 differences, 7, 325
 differentiating, 95

R

radians, 10, 12
radius, 12
 of convergence, 189
 mapping, 20
 trigonometric functions, 10
ranges
 sets, 3

trigonometric function values, 20
rates
 of change, 53–56, 64
 problem examples, 90–91
rational functions, 51
ratios, 181
 trigonometric functions, 10
Ratio Tests, 185–186
r-component, 34
reciprocal functions, 89
rectangles
 areas, 125
 coordinate systems, 33, 224
 regions, 141
 volume, 149
regions
 linearizing, 59
 rectangles, 141
relations, 3
 trigonometric formulas and, 16
relationships
 between coordinate systems, 33
 functions, 25–26
 between trigonometric/exponential functions, 25–26
representations
 graphs, 247–250
 partial derivatives, 243–246
 of products, 117
revolutions
 angular rate of, 21
 volume of, 146
Riemann sums, 126, 128
right-hand screw rule, 211
roots, 112
Root Test, 188
rotation fields, 273
rows
 multiplication, 207
 vectors, 196
rules, 3
 chain, 69, 86–90
 chain, partial derivatives, 246–247
 Cramer's, 215–216, 220
 differentiating, 102–103
 midpoint, 128
 products, 82–85, 162
 quotients, 85–86
 right-hand screw, 211
 Simpson's, 130
 trapezoids, 129
Runge-Kutta method, 323

S

saddle points, 263
saddle-shaped surfaces, 32
scalars, 195, 208–211,
   267–268
   gradients, 256
   multiplication, 206
secant, 10, 17
   derivatives, 95
   hyperbolic, 28
   lines, 60
second derivatives, 81–82,
   103–109, 260
second-order linear differential equations, 312–314
second power, functions
   raised to, 85
sections of curves, 140
separable equations, 309
separation of variables, 325
sequences, 114, 179–182
series, 179–182
   alternating, 186
   differential equations,
      317–318
   expanding, 188–194
   geometric, 193
   Harmonic Series, 185
   infinite, 183–184, 188
   Maclaurin, 188–189, 191
   powers, 188, 190, 318
   Taylor, 188–189, 191, 253
   tests, 187–188
sets
   domain, 4
   range, 3
sigma (θ) notation, 114–117
Simpson's rule, 130
sine, 10, 17, 51
   addition formulas for, 15
   curves, 24, 92
   hyperbolic, 27
   inverse, 98
   slopes, 24
   wave patterns, 23
slicing surfaces, 31
slopes, 24
   curves, approximating,
      109–112
   of cycloid, 234
   derivatives, 60–63
   directional derivatives, 258
   fields, 119
   functions, 6
   points, 57
   secant lines, 60
   of tangent lines, 58–60
solutions, 308
specified conditions, 308

speed, 54
   of particles, 238
spheres, 227
   coordinates, 36
   volume, 147, 153
spiral of Archimedes, 35
splitting integrals, 123
squares, matrices, 203
Stoke's Theorem, 277,
   304–306
straight lines, 62
   motion in, 230
stream functions, 287
strips, 127
subsidiary equations, 321
substitution integrals,
   164–172
subtraction
   complex numbers, 37
   formulas, 10, 14
   of functions, 5
   matrices, 205
   vectors, 199–202
sums
   of areas, 125
   associated errors, 128–131
   differentiating, 80–81
   of infinite series, 49
   integrals, 134–136
   of integrals, 136
   limits of, 116
   properties of, 115
   Riemann, 126, 128
   and sigma (θ) notation,
      114–117
   of vectors, 200
superposition, 313
surfaces, 30, 224–229
   integrals, 287–295
   saddle-shaped, 32
switching limits, 135
symmetric matrices, 204
symmetry, axis of, 40
systems of differential equations, 319–321

T

tables, 4
tangents, 10
   derivatives, 95
   hyperbolic, 28
   lines, 53, 58–60
   lines, equations, 109
   planes, 261
   vectors, 237
   velocity, 22
Taylor series, 188–189, 191,
   253
terms, 179

tests
   Comparison Tests,
      123–124, 184–185
   convergence, 183–184
   Integral Test, 187
   Ratio Tests, 185–186
   Root Test, 188
   series, 187–188
   vertical line, 5
three-dimensional space, 13
three-variable function, 30
time rate of change, 54
torque, 214
total volume, 149. *See also*
   volume
transposing matrices, 197,
   204, 219
trapezoid rules, 129
triangles
   matrices, 220
   trigonometric functions,
      10–11
   wedge sections, 35
trigonometric functions, 3,
   10–20
   differentiating, 91–95
   expansions of, 38
   inverse, 19
   inverse, derivatives, 95–99
   relationships, 25–26
triple integrals, 154
turning directions, 238
two-variable functions, 30,
   141
types
   of functions, 3–8
   of vectors, 196

U

units
   tangent vectors, 238
   vectors, 196–197, 210

V

values, 4
   averages, integrals, 131
   of cross products, 213
   trigonometric functions, 20
variables, 8, 243
   changing, 116, 152–157
   constants, 66–67
   derivatives, 66–67
   dummy, 114
   functions with multiple,
      29–33
   separation of, 325
   two-variable functions, 141
vectors, 267–268
   acceleration, 236, 241

addition and subtraction,
    199–202
columns, 220
conservative fields,
    276–282
definitions, 195–199
displacement, 235
divergence, 295
fields, 271–276
multiplication, 205–208
normal, 238
position, 224–229, 237,
    273
products, 211–215
velocity, 54, 230–241
    acceleration, 119
    angular, 21
    averages, 57
    derivatives, 60–63
    distances, increasing, 61
    fields, 275
    instantaneous, 57
    integrals, 124
    of oscillatory motion, 23

particles, 235
second derivatives, 81
tangents, 22
vectors, 196
vertical line tests, 5
vertices, 45
    points, 39
volume
    cubes, 215
    integrals, 145–152
    of paralleledpiped, 214
    rectangles, 149
    spheres, 153

W

waveforms, 12
wave patterns, 23
work, 157–159, 210
    along a curve, 279

X

x, derivatives of, 71–77
X-axis, 4, 8
    areas below, 131–134

derivatives, 71
integrals, 119
intervals, 148
XY-planes, 141

Y

Y-axis, 4, 8

Z

Z-axis, 34
    distance measurements,
        227
    flux, 289
    integrals, 147
z-component, 34
zeroes
    convergence, 49
    denominators, 28
    derivatives, 104
    derivatives of constants, 64
    divergence, 297
    limits, 48
    vectors, 197